Barbara O'Neill
Authentic Natural Remedies Collection

Barbara O. Mendez

© **Copyright 2024 - All rights reserved.**

The content contained within this book may not be reproduced, duplicated or transmitted without direct written permission from the author or the publisher.

Under no circumstances will any blame or legal responsibility be held against the publisher, or author, for any damages, reparation, or monetary loss due to the information contained within this book. Either directly or indirectly.

Legal Notice:

This book is copyright protected. This book is only for personal use. You cannot amend, distribute, sell, use, quote or paraphrase any part, or the content within this book, without the consent of the author or publisher.

Disclaimer Notice:

Please note the information contained within this document is for educational and entertainment purposes only. All effort has been executed to present accurate, up to date, and reliable, complete information. No warranties of any kind are declared or implied. Readers acknowledge that the author is not engaging in the rendering of legal, financial, medical or professional advice. The content within this book has been derived from various sources. Please consult a licensed professional before attempting any techniques outlined in this book.

By reading this document, the reader agrees that under no circumstances is the author responsible for any losses, direct or indirect, which are incurred as a result of the use of information contained within this document, including, but not limited to, — errors, omissions, or inaccuracies.

Introduction	**12**
Natural Remedies for Heart Health	**14**
Wholesome Harvest Heartbeet Salad	15
Soul-Soothing Lentil and Vegetable Stew	15
Revitalizing Salmon with Lemon-Dill Zest	16
Hearty Turkey Chili for a Happy Heart	16
Zesty Black Bean Soup with a Kick	17
Revitalizing Salmon with Garden-Fresh Dill Sauce	18
Light & Lively Turkey Lettuce Wraps	19
Nourishing Quinoa-Stuffed Bell Peppers	19
Creamy Greek Yogurt Chicken Salad	20
Mediterranean Tuna Salad with a Twist	21
Berry Blissful Oatmeal for a Heartfelt Morning	21
Avocado Toast with Egg-cellent Protein	22
Lemon Herb Grilled Chicken for a Zesty Zing	22
Sweet Potato Black Bean Fiesta Tacos	23
Vibrant Mediterranean Chickpea Salad	24
Flavorful Chicken Stir-Fry with Wholesome Brown Rice	24
Veggie-licious Chili for Heartfelt Warmth	25
Berrylicious Parfait with Heart-Healthy Granola	26
Spicy Tuna Salad with a Mediterranean Flair	26
Savory Lentil Soup with a Hint of Spice	27
Dietary and Lifestyle Tips for a Healthy Heart	29
Natural Remedies for Weight Management	**30**
Zesty Ginger-Lime Salmon with Mango Salsa	31
Metabolism-Boosting Turkey and Veggie Chili	31
Energizing Quinoa Breakfast Bowl with Berries and Nuts	32
Spicy Black Bean and Sweet Potato Power Bowl	33
Revitalizing Green Smoothie with Spinach and Pineapple	34
Aromatic Lentil Soup with Turmeric and Ginger	34
Colorful Salad with Avocado Dressing	35
Fire-Roasted Chicken with Metabolism-Boosting Spices	36
Protein-Packed Egg Muffins with Spinach and Feta	36
Whimsical Watermelon and Feta Salad with Mint	37
Sizzling Shrimp Stir-Fry with Zesty Ginger Sauce	37
Wholesome Turkey Lettuce Wraps with Peanut Sauce	38
Vibrant Veggie Omelet with a Kick of Cayenne	39
Creamy Avocado and Chicken Salad on Whole-Wheat Bread	40
Spicy Black Bean Burgers with Avocado Crema	40
Refreshing Cucumber and Tomato Salad with Dill Dressing	41
Metabolism-Boosting Chicken Soup with Ginger and Turmeric	42
Spicy Tuna Salad with a Mediterranean Twist	43
Sweet Potato and Black Bean Stuffed Peppers	43
Detoxifying Green Smoothie with Kale, Apple, and Ginger	44
Useful tips for weight management	45
Natural Remedies for Blood Sugar Balance	**46**
Savory Salmon Cakes with Zesty Dill Yogurt Sauce	47
Harvest Veggie and Lentil Power Bowl with Lemon-Tahini Dressing	47
Spiced Chickpea and Sweet Potato Curry with Coconut Milk	48
Hearty Black Bean Soup with Avocado and Lime Crema	49

Flavorful Chicken Fajita Salad with Avocado-Lime Ranch   50

Spicy Shrimp and Vegetable Stir-Fry with Brown Rice   50

Turkey Meatballs with Zucchini Noodles and Marinara Sauce   51

Baked Cod with Roasted Brussels Sprouts and Lemon-Garlic Drizzle   52

Spinach and Feta Stuffed Chicken Breast with Balsamic Glaze   53

Slow-Cooker Beef Stew with Root Vegetables and Herbs   53

One-Pan Roasted Chicken with Rainbow Vegetables and Herbs de Provence   54

Mediterranean Quinoa Salad with Chickpeas, Cucumber, and Feta   55

Egg White Frittata with Spinach, Mushrooms, and Goat Cheese   55

Chia Seed Pudding with Mixed Berries and Almonds   56

Avocado Toast with Smoked Salmon and Everything Bagel Seasoning   57

Spicy Black Bean Burgers with Mango Salsa and Guacamole   57

Turkey Chili with Butternut Squash and Kidney Beans   58

Greek Yogurt Parfait with Berries, Granola, and Honey   59

Calories Chicken Breast with Balsamic Glaze   59

Cauliflower Rice Stir-Fry with Chicken and Peanut Sauce   60

Lemony Lentil Soup with Spinach and Turmeric   61
    Useful tips for Blood Sugar Balance   62

Natural Remedies for Joint Health   63

Turmeric-Ginger Anti-Inflammatory Smoothie   64

Omega-3 Rich Salmon and Avocado Salad   64

Bone Broth with Fresh Herbs and Garlic   65

Anti-Inflammatory Sweet Potato and Quinoa Bowl   66

Curcumin-Infused Golden Milk Latte   66

Green Tea and Berry Antioxidant Smoothie   67

Collagen-Boosting Chicken Soup with Vegetables   67

Anti-Inflammatory Turmeric and Coconut Lentil Stew   68

Pineapple and Chia Seed Anti-Inflammatory Parfait   69

Spicy Ginger and Lemongrass Shrimp Stir-Fry   69

Omega-3 Packed Walnut and Flaxseed Porridge   70

Tart Cherry and Kale Anti-Inflammatory Salad   71

Zesty Lemon and Garlic Broccoli Stir-Fry   71

Green Smoothie with Spinach, Avocado, and Flaxseeds   72

Anti-Inflammatory Baked Turmeric Cauliflower   72

Ginger and Miso Soup with Seaweed   73

Spicy Turmeric and Lentil Stew   74

Garlic-Roasted Brussels Sprouts with Walnuts   74

Cucumber and Turmeric Detox Water   75

Anti-Inflammatory Blueberry and Almond Oatmeal   75
    Main Anti-Inflammatory and Pain-Relieving Herbs   77

Nutritional Strategies for Optimal Eye Health   78

Zesty Kale and Quinoa Salad with Lemon-Tahini Dressing for Eye-Brightening Vitamin A   79

Salmon Power Bowl with Sweet Potato and Spinach for Omega-3 Rich Vision Support   79

Sunshine Smoothie with Mango, Carrot, and Orange for Antioxidant-Packed Eye Protection   80

Berry Blissful Parfait with Greek Yogurt and Nuts for a Lutein and Zeaxanthin Boost   81

Savory Spinach and Feta Frittata for a Vision-Enhancing Breakfast   81

Hearty Lentil Soup with Carrots and Kale for a Zinc-Rich Eye Health Meal ... 82

One-Pan Roasted Chicken with Brussels Sprouts and Sweet Potatoes for a Vitamin-Packed Dinner ... 83

Vibrant Bell Pepper and Black Bean Salad with Avocado for Antioxidant-Rich Eye Fuel ... 84

Flavorful Turkey and Sweet Potato Chili with Spinach for a Vision-Supporting Feast ... 84

Nourishing Sweet Potato and Black Bean Burgers with Avocado Crema for Eye Health ... 85

Revitalizing Carrot and Ginger Soup with Coconut Milk for a Vitamin A-Rich Delight ... 86

Eye-Opening Berry and Nut Oatmeal with Chia Seeds for Omega-3 and Fiber ... 87

Colorful Rainbow Salad with Mixed Greens, Bell Peppers, and Sunflower Seeds for Antioxidant Power ... 87

Satisfying Salmon Cakes with Dill Yogurt Sauce for Omega-3 and Protein ... 88

Energizing Trail Mix with Nuts, Seeds, and Dried Fruit for Eye-Healthy Snacking ... 89

Refreshing Watermelon and Feta Salad with Mint for a Hydrating and Nutrient-Rich Treat ... 89

Wholesome Chicken Stir-Fry with Broccoli and Carrots for a Vision-Boosting Meal ... 90

Creamy Avocado and Spinach Smoothie with Banana for a Lutein-Rich Boost ... 91

Spicy Black Bean and Corn Salad with Cilantro Lime Dressing for Antioxidant Protection ... 91

Comforting Chicken Noodle Soup with Carrots and Celery for a Nourishing and Hydrating Meal ... 92

Foods Rich in Antioxidants for Eye Health ... 94
Eye Exercises and Relaxation Techniques ... 94

Natural Hair Care Recipes for Healthy, Strong, and Radiant Hair ... 96

Aloe Vera and Coconut Oil Deep Conditioning Hair Mask ... 97

Avocado and Honey Nourishing Hair Treatment ... 97

Apple Cider Vinegar Rinse for Shiny Hair ... 97

Rosemary and Lavender Hair Growth Serum ... 98

Banana and Olive Oil Moisturizing Hair Mask ... 98

Yogurt and Egg Protein Treatment for Stronger Hair ... 98

Green Tea and Peppermint Scalp Detox ... 99

Castor Oil and Jojoba Oil Split Ends Repair Serum ... 99

Chamomile and Lemon Lightening Hair Rinse ... 100

Flaxseed Gel for Natural Hair Styling ... 100

Henna and Amla Strengthening Hair Pack ... 100

Hibiscus and Coconut Milk Hair Thickening Mask ... 101

Onion Juice and Garlic Anti-Hair Loss Treatment ... 101

Shea Butter and Argan Oil Frizz Control Cream ... 102

Baking Soda and Water Clarifying Shampoo ... 102

Aloe Vera and Neem Scalp Soothing Gel ... 102

Oatmeal and Almond Milk Hydrating Hair Mask ... 103

Lemon Juice and Honey Anti-Dandruff Treatment ... 103

Fenugreek and Yogurt Hair Growth Mask ... 103

Rice Water Rinse for Strengthening Hair ... 104
Tips for Proper Scalp Hygiene ... 105

Natural Oral Health Recipes for Healthy Teeth and Gums ... 106

Baking Soda and Coconut Oil Whitening Toothpaste ... 107

Aloe Vera and Peppermint Mouthwash ... 107

Turmeric and Coconut Oil Gum Healing Paste ... 107

Clove Oil and Sea Salt Sore Gum Relief Rinse ... 108

Activated Charcoal and Bentonite Clay Detox Tooth Powder ... 108

Green Tea and Xylitol Fresh Breath Mouthwash ... 108

Myrrh and Echinacea Anti-Inflammatory Gum Gel 109

Cranberry and Vitamin C Plaque-Reducing Mouth Rinse 109

Olive Oil and Lemon Oil Pulling Solution 109

Neem and Tea Tree Oil Antibacterial Mouthwash 110

Sage and Sea Salt Teeth Strengthening Rinse 110

Hydrogen Peroxide and Baking Soda Whitening Gel 111

Cinnamon and Honey Antibacterial Mouth Rinse 111

Propolis and Aloe Vera Healing Mouth Gel 111

Eucalyptus Oil and Peppermint Oil Fresh Breath Spray 112

Licorice Root and Spearmint Cavity Prevention Toothpaste 112

Chamomile and Clove Gum Soothing Rinse 112

Calcium Powder and Xylitol Remineralizing Tooth Powder 113

Basil and Fennel Natural Breath Freshener 113

Apple Cider Vinegar and Baking Soda Natural Tooth Cleaner 114

Tips for Proper Oral Hygiene 115

Natural Recipes for Women's Health 116

Hormone-Balancing Flaxseed and Berry Smoothie 117

Iron-Rich Beet and Spinach Salad 117

Anti-Inflammatory Turmeric and Ginger Tea 118

Calcium-Boosting Almond and Kale Smoothie 118

Menopause Relief Sage and Lemon Balm Tea 119

PMS Soothing Chamomile and Raspberry Leaf Tea 119

Bone Strengthening Sesame and Chia Seed Pudding 120

Skin-Glowing Avocado and Cucumber Salad 120

Stress-Relief Lavender and Valerian Root Tea 121

Energy-Boosting Maca and Banana Smoothie 122

Immune-Boosting Elderberry and Hibiscus Tea 122

Detoxifying Dandelion and Mint Tea 123

Mood-Enhancing Dark Chocolate and Walnut Bites 123

Digestive Health Fennel and Peppermint Tea 124

Uterine Health Red Clover and Nettle Infusion 124

Hair and Nail Strengthening Biotin-Rich Smoothie 125

Heart Health Omega-3 Flaxseed and Walnut Mix 125

Hydrating Coconut Water and Aloe Vera Drink 126

Anti-Aging Green Tea and Blueberry Smoothie 126

Reproductive Health Maca and Cinnamon Latte 127

Herbs for Menstrual Regulation, Menopause, and Fertility 128

30-hour video course 129

Natural Recipes for Men's Health 130

Testosterone-Boosting Walnut and Banana Smoothie 131

Heart-Healthy Avocado and Spinach Salad 131

Anti-Inflammatory Turmeric and Black Pepper Tea 132

Energy-Enhancing Maca and Almond Smoothie 132

Prostate Health Pomegranate and Flaxseed Juice 133

Muscle Recovery Tart Cherry and Ginger Smoothie 133

Immune-Boosting Elderberry and Echinacea Tea 134

Stress-Reducing Ashwagandha and Cinnamon Latte 134

Brain-Boosting Blueberry and Chia Seed Smoothie 135

Detoxifying Green Tea and Lemon Drink 135
Bone Strengthening Sesame and Almond Milk 136
Libido-Enhancing Pumpkin Seed and Cacao Smoothie 136
Anti-Anxiety Lavender and Chamomile Tea 137
Liver Health Beet and Carrot Juice 138
Digestive Aid Fennel and Peppermint Tea 138
Blood Pressure Regulating Hibiscus and Hawthorn Tea 139
Anti-Aging Dark Chocolate and Walnut Bites 139
Hair Growth Rosemary and Nettle Infusion 140
Skin-Clearing Aloe Vera and Lemon Drink 140
Joint Health Turmeric and Pineapple Smoothie 141
Herbs for Prostate Health, Fertility, and Libido 142

Natural Recipes for Kidney Health 143

Kidney Cleansing Watermelon and Mint Juice 144
Anti-Inflammatory Turmeric and Ginger Tea 144
Detoxifying Cranberry and Lemon Drink 145
Parsley and Celery Kidney Flush Smoothie 145
Dandelion and Nettle Detox Tea 146
Hydrating Cucumber and Lemon Water 146
Apple and Carrot Kidney Cleansing Juice 147
Potassium-Rich Banana and Spinach Smoothie 147
Anti-Oxidant Blueberry and Beet Juice 148
Hydrating Coconut Water and Aloe Vera Drink 148
Detox Green Apple and Kale Smoothie 149
Anti-Inflammatory Pineapple and Turmeric Juice 149
Lemon and Ginger Detox Water 150
Healing Aloe Vera and Cucumber Juice 150
Anti-Oxidant Rich Pomegranate and Beet Juice 151
Cleansing Celery and Apple Smoothie 151
Hydration Boosting Coconut and Lime Drink 152
Anti-Inflammatory Ginger and Lemon Tea 152
Detoxifying Green Tea and Cranberry Drink 153
Kidney Health Watermelon and Cucumber Juice 153
Dietary Tips for Kidney Health 155

Natural Recipes for Liver Health 156

Green Detox Smoothie with Kale and Lemon 157
Turmeric and Ginger Liver Cleanse Tea 157
Beetroot and Carrot Liver Tonic Juice 158
Avocado and Spinach Salad with Apple Cider Vinaigrette 158
Cilantro and Lime Liver Detox Soup 159
Dandelion Greens and Quinoa Liver Power Bowl 160
Lemon and Mint Liver Flush Water 160
Broccoli and Turmeric Stir-Fry with Brown Rice 161
Garlic and Olive Oil Liver Detox Dressing 162
Apple, Cucumber, and Celery Liver Cleanse Juice 162
Spirulina and Pineapple Liver Boost Smoothie 163
Asparagus and Lemon Zest Detox Salad 163
Chia Seed and Blueberry Liver Rejuvenation Pudding 164
Bitter Melon and Ginger Healing Broth 164
Red Cabbage and Beet Liver Support Slaw 165
Artichoke and Lemon Roasted Veggie Medley 166
Flaxseed and Papaya Liver Revitalizing Smoothie 166
Wheatgrass and Orange Morning Detox Juice 167
Parsley and Lemon Liver Cleansing Soup 168
Green Apple and Kale Liver Purification Salad 168
Tips for a Detoxifying Diet 170

Natural Recipes for Bone Health 171

Calcium-Rich Kale and Almond Smoothie 172

Bone-Strengthening Broccoli and Salmon Salad 172

Vitamin D-Fortified Mushroom and Spinach Omelette 173

Magnesium-Boosting Avocado and Quinoa Bowl 173

Bone-Building Sesame and Chickpea Hummus 174

Calcium-Loaded Orange and Fig Breakfast Parfait 175

Anti-Inflammatory Turmeric and Ginger Bone Broth 175

Bone-Healthy Sardine and Arugula Salad 176

Collagen-Rich Chicken and Vegetable Soup 177

Bone-Bolstering Tahini and Carrot Slaw 177

Calcium-Packed Almond and Berry Overnight Oats 178

Bone-Supporting Lentil and Spinach Stew 179

Magnesium-Enriched Sweet Potato and Black Bean Tacos 179

Omega-3 Rich Walnut and Blueberry Salad 180

Bone-Nourishing Butternut Squash and Kale Risotto 181

Calcium-Fortified Chia and Coconut Pudding 181

Bone-Beneficial Edamame and Sesame Stir-Fry 182

Phosphorus-Packed Pumpkin Seed and Avocado Toast 183

Bone-Strengthening Greek Yogurt and Almond Smoothie Bowl 183

Calcium-Enhanced Tofu and Bok Choy Stir-Fry 184

Foods Rich in Calcium and Vitamin D 185

Natural Recipes for Hormonal Balance for Women and Men 186

Hormone-Balancing Smoothie with Maca and Berries 187

Herbal Infusion for Female Hormone Harmony 187

Roasted Vegetables with Thyme and Sage for Thyroid Support 188

Flaxseed and Chia Pudding for Estrogen Balance 188

Pumpkin Seed Protein Balls for Testosterone Boost 189

Detoxifying Green Juice with Dandelion and Parsley 190

Spicy Lentil Soup with Turmeric and Ginger for Inflammation Reduction 190

Omega-3 Rich Salmon Salad with Avocado and Walnuts 191

Adaptogenic Herbal Tea with Ashwagandha and Holy Basil 192

Fermented Sauerkraut for Gut Health and Hormone Regulation 192

Sweet Potato and Spinach Frittata for Adrenal Support 193

Hormone-Regulating Golden Milk with Turmeric and Coconut Milk 194

Broccoli and Cauliflower Stir-Fry for Estrogen Detox 194

Bone Broth with Garlic and Rosemary for Immune Support 195

Quinoa Salad with Beets and Arugula for Liver Detoxification 196

Almond Butter Energy Bars with Seeds and Nuts 196

Coconut and Berry Parfait with Probiotic Yogurt 197

Hormone-Balancing Smoothie Bowl with Spirulina and Banana 197

Sesame Seed and Honey Bars for Hormone Support 198

Zucchini Noodles with Pesto for Anti-Inflammatory Benefits 199

Herbs for Stress Management and Hormonal Balance 200

Natural Recipes for Mental and Brain Health   201

Brain-Boosting Blueberry and Spinach Smoothie 202

Walnut and Berry Salad with Mixed Greens   202

Turmeric and Black Pepper Golden Milk for Cognitive Health   203

Omega-3 Rich Chia Seed Pudding with Almond Milk   203

Dark Chocolate and Almond Energy Bars for Mental Clarity   204

Roasted Beet and Goat Cheese Salad for Enhanced Memory   204

Avocado and Tomato Brain-Healthy Toast   205

Ginkgo Biloba Herbal Tea for Focus and Concentration   206

Salmon and Quinoa Power Bowl with Leafy Greens   206

Turmeric and Ginger Spiced Carrot Soup   207

Green Tea and Lemon Detox Drink for Mental Alertness   208

Flaxseed and Berry Yogurt Parfait for Brain Health   208

Rosemary and Olive Oil Roasted Sweet Potatoes   209

Coconut and Berry Smoothie Bowl for Cognitive Function   209

Pumpkin Seed and Spinach Pesto Pasta   210

Memory-Enhancing Ginseng and Honey Herbal Infusion   211

Dark Leafy Greens and Citrus Salad with Walnuts   211

Baked Salmon with Garlic and Rosemary   212

Nutty Quinoa and Spinach Salad for Mental Clarity   213

Cacao and Almond Brain-Boosting Smoothie   213
    Nootropic and Adaptogenic Herbs for Brain Health   215

Natural Recipes for Physical Recovery   216

Anti-Inflammatory Turmeric and Ginger Smoothie   217

Cherry and Almond Recovery Smoothie   217

Protein-Packed Quinoa and Black Bean Salad   218

Omega-3 Rich Salmon with Avocado Salsa   218

Baked Sweet Potato and Spinach Frittata   219

Rehydrating Coconut Water and Pineapple Drink   220

Muscle-Repairing Chicken and Broccoli Stir-Fry 220

Beetroot and Carrot Juice for Endurance   221

Protein-Packed Greek Yogurt with Berries and Nuts   222

Anti-Fatigue Green Tea and Lemon Cooler   222

Pumpkin Seed and Oat Energy Bites   223

Banana and Peanut Butter Recovery Smoothie 223

Hearty Lentil and Vegetable Stew   224

Chia Seed Pudding with Coconut Milk and Mango   225

Tropical Mango and Chia Recovery Drink   225

Walnut and Spinach Pesto Pasta   226

Hydrating Watermelon and Mint Salad   226

Anti-Inflammatory Golden Milk Latte   227

Blueberry and Kale Smoothie Bowl   228

Spicy Ginger and Turmeric Carrot Soup   228
    Natural Supplements and Foods for Athletes   230

Natural Recipes for Gut Health   231

Fermented Sauerkraut for Gut Health   232

Probiotic-Rich Kimchi   232

Kefir Smoothie with Berries and Honey   233

Homemade Kombucha Tea   234

Bone Broth with Ginger and Turmeric   234

Greek Yogurt and Flaxseed Parfait	235
Miso Soup with Tofu and Seaweed	235
Garlic and Herb Marinated Olives	236
Roasted Asparagus with Lemon and Garlic	237
Pumpkin and Ginger Soup	237
Avocado and Spinach Green Smoothie	238
Chia Seed Pudding with Coconut Milk	238
Apple Cider Vinegar Detox Drink	239
Beet and Carrot Salad with Apple Cider Vinegar Dressing	240
Quinoa Salad with Fermented Vegetables	240
Blueberry and Almond Overnight Oats	241
Cabbage and Carrot Coleslaw with Yogurt Dressing	241
Turmeric and Ginger Infused Water	242
Garlic and Rosemary Roasted Sweet Potatoes	243
Probiotic Banana and Oat Muffins	243
Probiotics, Prebiotics, and Fermented Foods for Gut Health	245

Conclusion   246

GET YOUR 6 BARBARA O'NEILL BONUSES

👉 SCAN HERE TO DOWNLOAD IT

Introduction

The pursuit of optimal health and wellness is a journey that touches every aspect of our lives. In today's fast-paced world, it's easy to overlook the profound impact that our daily choices have on our overall well-being. Dr. Barbara O'Neill, a renowned advocate for natural healing, emphasizes the importance of harnessing the power of nature to promote health and vitality. This book is a comprehensive guide designed to help you achieve better health through natural remedies and wholesome recipes, all rooted in the principles and teachings of Dr. O'Neill.

In these pages, you will find a wealth of information on natural remedies that support various aspects of health, from heart health and weight management to blood sugar balance and joint health. Each chapter offers practical advice, dietary tips, and delicious recipes that are not only easy to prepare but also deeply nourishing. The goal is to empower you with the knowledge and tools needed to make healthier choices and foster a more balanced, vibrant life.

Heart health, for instance, is critical to overall well-being. The recipes and tips provided in this section aim to strengthen your cardiovascular system, reduce inflammation, and promote a healthy heart. From Wholesome Harvest Heartbeet Salad to Revitalizing Salmon with Lemon-Dill Zest, these dishes are designed to support your heart health naturally.

Weight management is another crucial aspect of maintaining a healthy lifestyle. The natural remedies and recipes in this section focus on boosting metabolism, enhancing energy levels, and supporting weight loss through nutrient-dense foods. Discover delicious options like Zesty Ginger-Lime Salmon with Mango Salsa and Energizing Quinoa Breakfast Bowl with Berries and Nuts, which help you stay on track with your weight management goals.

Balancing blood sugar levels is essential for preventing chronic diseases such as diabetes. The recipes here, such as Savory Salmon Cakes with Zesty Dill Yogurt Sauce and Hearty Black Bean Soup with Avocado and Lime Crema, are crafted to help stabilize blood sugar levels and provide sustained energy throughout the day.

Joint health is also a significant concern for many, especially as we age. This book offers a variety of anti-inflammatory recipes and remedies, like Turmeric-Ginger Anti-Inflammatory Smoothie and Bone Broth with Fresh Herbs and Garlic, to support joint health and reduce pain naturally.

In addition to these focused sections, the book delves into eye health, hair care, oral health, and more. Each section provides targeted recipes and lifestyle tips to address specific health needs. For women and men, there are dedicated chapters that explore hormonal balance, reproductive health, and vitality, offering customized solutions for different stages of life.

Kidney and liver health are critical for detoxification and overall well-being. The recipes in these sections aim to cleanse and regenerate these vital organs, using natural ingredients known for their detoxifying properties. Similarly, the sections on bone health and mental clarity provide comprehensive strategies to strengthen bones and enhance cognitive function.

Gut health is foundational to overall wellness, as a healthy gut contributes to improved digestion, immunity, and even mental health. The natural recipes for gut health, including Fermented Sauerkraut and Kefir Smoothie with Berries and Honey, promote a balanced microbiome and better digestion.

By embracing the natural remedies and recipes outlined in this book, you are taking a significant step towards improving your health and well-being. Remember, the journey to optimal health is not about quick fixes but about making consistent, mindful choices that support your body's natural healing processes.

Thank you for choosing this book as your guide to natural health. May it inspire you to live a healthier, more balanced life, and may you find joy and vitality in every recipe and remedy you try. After you have explored these pages and experienced the benefits, I invite you to leave a review. Your feedback is invaluable and helps spread the word about natural health solutions. I truly hope you enjoy this book and find it as enriching to read as it was for me to create.

Natural Remedies for Heart Health

Wholesome Harvest Heartbeet Salad

Prep Time: 15 minutes
Portion Size: 2

Ingredients:
- 4 cups mixed greens (arugula, spinach, kale)
- 1 medium beet, roasted and cubed
- ½ cup crumbled goat cheese
- ¼ cup chopped walnuts
- 2 tablespoons balsamic vinaigrette

Instructions:
In a large bowl, combine greens, beets, goat cheese, and walnuts. Drizzle with balsamic vinaigrette and toss gently.

Nutritional Information (per serving):
Calories: 280
Carbohydrates: 15g
Protein: 12g
Fat: 18g

Soul-Soothing Lentil and Vegetable Stew

Prep Time: 10 minutes
Cook Time: 30 minutes
Portion Size: 4

Ingredients:
- 1 tablespoon olive oil
- 1 onion, chopped
- 2 cloves garlic, minced
- 2 carrots, chopped
- 2 stalks celery, chopped
- 1 cup brown lentils, rinsed
- 4 cups vegetable broth
- 1 teaspoon dried thyme
- ½ teaspoon salt
- ¼ teaspoon black pepper
- ½ cup chopped kale

Instructions:

Heat olive oil in a large pot over medium heat. Add onion, garlic, carrots, and celery. Cook until softened.
Add lentils, broth, thyme, salt, and pepper. Bring to a boil, then reduce heat and simmer for 20 minutes.
Add kale and cook for 5 minutes more, or until lentils are tender and kale is wilted.

Nutritional Information (per serving):
Calories: 260
Carbohydrates: 35g
Protein: 16g
Fat: 7g

Revitalizing Salmon with Lemon-Dill Zest

Prep Time: 5 minutes
Cook Time: 15 minutes
Portion Size: 2

Ingredients:
2 salmon fillets (6 ounces each)
1 tablespoon olive oil
1 lemon, zested and juiced
¼ cup chopped fresh dill
Salt and pepper to taste

Instructions:
Preheat oven to 400°F (200°C).
Place salmon fillets on a baking sheet lined with parchment paper.
Drizzle with olive oil and season with salt and pepper.
Top with lemon zest and dill.
Bake for 12-15 minutes, or until salmon is cooked through.
Squeeze lemon juice over salmon before serving.

Nutritional Information (per serving):
Calories: 280
Carbohydrates: 5g
Protein: 30g
Fat: 15g

Hearty Turkey Chili for a Happy Heart

Prep Time: 15 minutes
Cook Time: 45 minutes

Portion Size: 6

Ingredients:
- 1 tablespoon olive oil
- 1 onion, chopped
- 2 cloves garlic, minced
- 1 pound ground turkey
- 1 red bell pepper, chopped
- 1 can (15 ounces) diced tomatoes, undrained
- 1 can (15 ounces) kidney beans, rinsed and drained
- 1 can (15 ounces) black beans, rinsed and drained
- 1 cup corn kernels
- 1 teaspoon chili powder
- 1/2 teaspoon cumin
- 1/4 teaspoon salt
- 1/4 teaspoon black pepper

Instructions:
1. Heat olive oil in a large pot over medium heat. Add onion, garlic, and bell pepper. Cook until softened.
2. Add ground turkey and cook until browned. Drain excess grease.
3. Stir in tomatoes, kidney beans, black beans, corn, chili powder, cumin, salt, and pepper.
4. Bring to a boil, then reduce heat and simmer for 30 minutes.
5. Serve with your favorite toppings, such as avocado, chopped cilantro, or Greek yogurt.

Nutritional Information (per serving):
- Calories: 350
- Carbohydrates: 40g
- Protein: 25g
- Fat: 10g

Zesty Black Bean Soup with a Kick

Prep Time: 10 minutes
Cook Time: 30 minutes
Portion Size: 4

Ingredients:
- 1 tablespoon olive oil
- 1 onion, chopped
- 2 cloves garlic, minced
- 1 jalapeno pepper, seeded and minced
- 1 teaspoon chili powder
- 1/2 teaspoon cumin
- 1/4 teaspoon smoked paprika

- 1/4 teaspoon salt
- 1/4 teaspoon black pepper
- 2 cans (15 ounces) black beans, rinsed and drained
- 4 cups vegetable broth
- 1/4 cup chopped cilantro
- Lime wedges, for serving

Instructions:
1. Heat olive oil in a large pot over medium heat. Add onion, garlic, and jalapeno. Cook until softened.
2. Stir in chili powder, cumin, smoked paprika, salt, and pepper. Cook for 1 minute more.
3. Add black beans and broth. Bring to a boil, then reduce heat and simmer for 20 minutes.
4. Serve with a sprinkle of cilantro and a squeeze of lime.

Nutritional Information (per serving):
- Calories: 250
- Carbohydrates: 35g
- Protein: 15g
- Fat: 7g

Revitalizing Salmon with Garden-Fresh Dill Sauce

Prep Time: 10 minutes
Cook Time: 15 minutes
Portion Size: 2

Ingredients:
- 2 salmon fillets (6 ounces each)
- 1 tablespoon olive oil
- 1/4 cup plain Greek yogurt
- 1/4 cup chopped fresh dill
- 1 tablespoon lemon juice
- 1/4 teaspoon salt
- 1/4 teaspoon black pepper

Instructions:
1. Preheat oven to 400°F (200°C).
2. Place salmon fillets on a baking sheet lined with parchment paper.
3. Drizzle with olive oil and season with salt and pepper.
4. Bake for 12-15 minutes, or until salmon is cooked through.
5. While salmon is baking, combine yogurt, dill, lemon juice, salt, and pepper in a small bowl.
6. Serve salmon topped with dill sauce.

Nutritional Information (per serving):
- Calories: 300

- Carbohydrates: 5g
- Protein: 30g
- Fat: 18g

Light & Lively Turkey Lettuce Wraps

Prep Time: 15 minutes
Cook Time: 10 minutes
Portion Size: 4

Ingredients:
- 1 pound ground turkey
- 1 tablespoon olive oil
- 1/2 cup chopped onion
- 2 cloves garlic, minced
- 1/2 cup shredded carrots
- 1/4 cup chopped water chestnuts
- 2 tablespoons hoisin sauce
- 2 tablespoons soy sauce
- 1 tablespoon rice vinegar
- 1 teaspoon sesame oil
- 1 head butter lettuce, leaves separated

Instructions:
1. Heat olive oil in a large skillet over medium heat. Add turkey and cook until browned.
2. Add onion, garlic, carrots, and water chestnuts. Cook until softened.
3. Stir in hoisin sauce, soy sauce, rice vinegar, and sesame oil. Cook for 1 minute more.
4. Spoon turkey mixture into lettuce leaves.

Nutritional Information (per serving):
- Calories: 250
- Carbohydrates: 10g
- Protein: 25g
- Fat: 12g

Nourishing Quinoa-Stuffed Bell Peppers

Prep Time: 15 minutes
Cook Time: 40 minutes
Portion Size: 4

Ingredients:

- 4 bell peppers (any color), halved and seeded
- 1 cup quinoa, cooked
- 1 can (15 ounces) black beans, rinsed and drained
- 1/2 cup corn kernels
- 1/2 cup chopped onion
- 1/4 cup chopped fresh cilantro
- 1/2 cup salsa
- 1/4 cup shredded cheddar cheese (optional)

Instructions:
1. Preheat oven to 375°F (190°C).
2. In a large bowl, combine quinoa, black beans, corn, onion, and cilantro.
3. Stuff bell pepper halves with the quinoa mixture.
4. Place in a baking dish and bake for 30 minutes.
5. Top with salsa and cheese (if using) and bake for an additional 10 minutes, or until cheese is melted and bubbly.

Nutritional Information (per serving):
- Calories: 350
- Carbohydrates: 45g
- Protein: 15g
- Fat: 10g

Creamy Greek Yogurt Chicken Salad

Prep Time: 10 minutes
Portion Size: 4

Ingredients:
- 2 cups cooked chicken, shredded
- 1 cup plain Greek yogurt
- 1/2 cup chopped celery
- 1/4 cup chopped red onion
- 1/4 cup chopped walnuts or pecans
- 2 tablespoons chopped fresh dill
- 1 tablespoon Dijon mustard
- Salt and pepper to taste
- Whole wheat bread, lettuce leaves, or crackers for serving

Instructionso:
1. In a large bowl, combine chicken, yogurt, celery, onion, walnuts, dill, and mustard.
2. Season with salt and pepper to taste.
3. Serve on whole wheat bread, lettuce leaves, or crackers.

Nutritional Information (per serving):
- Calories: 280

- Carbohydrates: 10g
- Protein: 30g
- Fat: 12g

Mediterranean Tuna Salad with a Twist

Prep Time: 15 minutes
Portion Size: 2

Ingredients:
- 1 can (5 ounces) tuna in olive oil, drained
- 1/2 cup chopped cucumber
- 1/4 cup chopped red onion
- 1/4 cup chopped Kalamata olives
- 2 tablespoons chopped fresh parsley
- 2 tablespoons capers, drained
- 1 tablespoon lemon juice
- Salt and pepper to taste
- Whole wheat bread or lettuce leaves for serving

Instructions:
1. In a medium bowl, combine all ingredients.
2. Season with salt and pepper.
3. Serve on whole wheat bread or lettuce leaves.

Nutritional Information (per serving):
- Calories: 300
- Carbohydrates: 10g
- Protein: 25g
- Fat: 18g

Berry Blissful Oatmeal for a Heartfelt Morning

Prep Time: 5 minutes
Cook Time: 5 minutes
Portion Size: 2

Ingredients:
- 1 cup rolled oats
- 2 cups unsweetened almond milk
- 1/2 cup mixed berries (blueberries, raspberries, strawberries)
- 1 tablespoon chia seeds

- 1 tablespoon chopped walnuts
- Drizzle of honey or maple syrup (optional)

Instructions:
1. Combine oats and almond milk in a saucepan.
2. Bring to a boil, then reduce heat and simmer for 5 minutes, or until thickened.
3. Stir in berries, chia seeds, and walnuts.
4. Sweeten with honey or maple syrup, if desired.

Nutritional Information (per serving):
- Calories: 250
- Carbohydrates: 40g
- Protein: 10g
- Fat: 8g

Avocado Toast with Egg-cellent Protein

Prep Time: 5 minutes
Portion Size: 1

Ingredients:
- 1 slice whole-grain bread, toasted
- ½ avocado, mashed
- 1 hard-boiled egg, sliced
- Pinch of red pepper flakes
- Salt and pepper to taste

Instructions:
1. Spread mashed avocado on toast.
2. Top with sliced egg.
3. Sprinkle with red pepper flakes, salt, and pepper.

Nutritional Information (per serving):
- Calories: 300
- Carbohydrates: 20g
- Protein: 15g
- Fat: 18g

Lemon Herb Grilled Chicken for a Zesty Zing

Prep Time: 10 minutes (plus marinating time)
Cook Time: 20 minutes

Portion Size: 2

Ingredients:
- 2 boneless, skinless chicken breasts
- 2 tablespoons olive oil
- 2 tablespoons lemon juice
- 1 tablespoon chopped fresh herbs (thyme, rosemary, oregano)
- ½ teaspoon salt
- ¼ teaspoon black pepper

Instructions:
1. In a small bowl, whisk together olive oil, lemon juice, herbs, salt, and pepper.
2. Place chicken breasts in a shallow dish and pour marinade over them. Marinate for at least 30 minutes or up to 4 hours in the refrigerator.
3. Preheat grill to medium heat.
4. Grill chicken for 6-8 minutes per side, or until cooked through (internal temperature of 165°F).

Nutritional Information (per serving):
- Calories: 250
- Carbohydrates: 5g
- Protein: 30g
- Fat: 12g

Sweet Potato Black Bean Fiesta Tacos

Prep Time: 10 minutes
Cook Time: 20 minutes
Portion Size: 4

Ingredients:
- 2 medium sweet potatoes, diced
- 1 tablespoon olive oil
- 1 teaspoon chili powder
- ½ teaspoon cumin
- ¼ teaspoon salt
- ¼ teaspoon black pepper
- 1 can (15 ounces) black beans, rinsed and drained
- 8 corn tortillas
- Toppings: chopped cilantro, diced avocado, salsa, lime wedges

Instructions:
1. Preheat oven to 400°F (200°C).
2. Toss sweet potatoes with olive oil, chili powder, cumin, salt, and pepper.
3. Spread on a baking sheet and roast for 20 minutes, or until tender.
4. Warm tortillas in a dry skillet or microwave.
5. Fill tortillas with sweet potatoes, black beans, and desired toppings.

Nutritional Information (per serving):
- Calories: 300
- Carbohydrates: 45g
- Protein: 10g
- Fat: 8g

Vibrant Mediterranean Chickpea Salad

Prep Time: 15 minutes
Portion Size: 4

Ingredients:
- 2 cans (15 ounces) chickpeas, rinsed and drained
- 1 cucumber, diced
- ½ red onion, diced
- ½ cup chopped fresh parsley
- ¼ cup chopped fresh mint
- ¼ cup crumbled feta cheese
- ¼ cup olive oil
- 2 tablespoons lemon juice
- Salt and pepper to taste

Instructions:
1. In a large bowl, combine all ingredients.
2. Toss to coat evenly.
3. Season with salt and pepper.

Nutritional Information (per serving):
- Calories: 320
- Carbohydrates: 35g
- Protein: 12g
- Fat: 15g

Flavorful Chicken Stir-Fry with Wholesome Brown Rice

Prep Time: 15 minutes
Cook Time: 20 minutes
Portion Size: 4

Ingredients:
- 1 tablespoon olive oil
- 1 pound boneless, skinless chicken breasts, cut into strips

- 1 onion, sliced
- 2 cloves garlic, minced
- 1 red bell pepper, sliced
- 1 green bell pepper, sliced
- 1 cup broccoli florets
- ½ cup low-sodium soy sauce
- ¼ cup honey
- 1 tablespoon cornstarch
- 4 cups cooked brown rice

Instructions:
1. Heat olive oil in a large skillet or wok over medium-high heat. Add chicken and cook until browned.
2. Add onion, garlic, and bell peppers. Cook until softened.
3. Add broccoli and cook until tender-crisp.
4. In a small bowl, whisk together soy sauce, honey, and cornstarch. Add to skillet and cook until sauce thickens.
5. Serve over brown rice.

Nutritional Information (per serving):
- Calories: 450
- Carbohydrates: 50g
- Protein: 30g
- Fat: 12g

Veggie-licious Chili for Heartfelt Warmth

Prep Time: 15 minutes
Cook Time: 45 minutes
Portion Size: 6

Ingredients:
- 1 tablespoon olive oil
- 1 onion, chopped
- 2 cloves garlic, minced
- 1 green bell pepper, chopped
- 1 red bell pepper, chopped
- 2 zucchini, diced
- 1 can (15 ounces) kidney beans, rinsed and drained
- 1 can (15 ounces) black beans, rinsed and drained
- 1 can (14.5 ounces) diced tomatoes, undrained
- 1 can (4 ounces) diced green chilies
- 1 teaspoon chili powder
- 1/2 teaspoon cumin

- 1/4 teaspoon salt
- 1/4 teaspoon black pepper

Instructions:
1. Heat olive oil in a large pot over medium heat. Add onion, garlic, and bell peppers. Cook until softened.
2. Add zucchini and cook until tender.
3. Stir in kidney beans, black beans, tomatoes, green chilies, chili powder, cumin, salt, and pepper.
4. Bring to a boil, then reduce heat and simmer for 30 minutes.
5. Serve with your favorite toppings, such as avocado, chopped cilantro, or Greek yogurt.

Nutritional Information (per serving):
- Calories: 300
- Carbohydrates: 40g
- Protein: 18g
- Fat: 8g

Berrylicious Parfait with Heart-Healthy Granola

Prep Time: 5 minutes
Portion Size: 2

Ingredients:
- 1 cup plain Greek yogurt
- 1/2 cup mixed berries (blueberries, raspberries, strawberries)
- 1/4 cup granola
- 1 tablespoon honey (optional)

Instructions:
1. Layer yogurt, berries, and granola in two glasses or jars.
2. Drizzle with honey, if desired.

Nutritional Information (per serving):
- Calories: 200
- Carbohydrates: 25g
- Protein: 12g
- Fat: 5g

Spicy Tuna Salad with a Mediterranean Flair

Prep Time: 10 minutes
Portion Size: 2

Ingredients:

- 1 can (5 ounces) tuna in water, drained
- 1/4 cup chopped red onion
- 1/4 cup chopped Kalamata olives
- 2 tablespoons chopped sun-dried tomatoes
- 1 tablespoon capers, drained
- 2 tablespoons olive oil
- 1 tablespoon lemon juice
- 1/4 teaspoon red pepper flakes
- Salt and pepper to taste
- Whole wheat bread or lettuce leaves for serving

Instructions:
1. In a medium bowl, combine all ingredients.
2. Season with salt and pepper.
3. Serve on whole wheat bread or lettuce leaves.

Nutritional Information (per serving):
- Calories: 300
- Carbohydrates: 10g
- Protein: 25g
- Fat: 18g

Savory Lentil Soup with a Hint of Spice

Prep Time: 10 minutes
Cook Time: 30 minutes
Portion Size: 4

Ingredients:
- 1 tablespoon olive oil
- 1 onion, chopped
- 2 cloves garlic, minced
- 2 carrots, chopped
- 2 stalks celery, chopped
- 1 cup green lentils, rinsed
- 4 cups vegetable broth
- 1 teaspoon curry powder
- ½ teaspoon turmeric
- ¼ teaspoon salt
- ¼ teaspoon black pepper
- ¼ cup chopped cilantro for garnish

Instructions:
1. Heat olive oil in a large pot over medium heat. Add onion, garlic, carrots, and celery. Cook until softened.

2. Add lentils, broth, curry powder, turmeric, salt, and pepper. Bring to a boil, then reduce heat and simmer for 25 minutes, or until lentils are tender.
3. Serve garnished with cilantro.

Nutritional Information (per serving):
- Calories: 270
- Carbohydrates: 40g
- Protein: 15g
- Fat: 7g

Dietary and Lifestyle Tips for a Healthy Heart

1. **Choose whole grains over refined grains**: Whole grains, such as oats, quinoa, and brown rice, are rich in fiber, vitamins, and essential minerals. These nutrients help regulate blood sugar levels, improve digestion, and reduce the risk of heart disease.
2. **Limit saturated and trans fats**: Saturated fats, found in foods like butter, red meat, and full-fat dairy products, and trans fats, often found in baked goods and fried foods, can raise LDL (bad) cholesterol levels. Reducing the intake of these fats is essential to keep arteries clear and the heart healthy.
3. **Prioritize healthy fats**: Replacing saturated fats with monounsaturated and polyunsaturated fats, such as those found in olive oil, avocados, nuts, seeds, and fatty fish (like salmon and sardines), can help improve blood lipid profiles and reduce inflammation.
4. **Eat plenty of fruits and vegetables**: Fruits and vegetables are rich in antioxidants, fiber, vitamins, and minerals that support heart health. Consuming at least five servings a day can help lower blood pressure, improve vascular function, and reduce the risk of heart disease.
5. **Reduce sodium intake**: Excessive sodium consumption can raise blood pressure, a major risk factor for heart disease. It is advisable to limit the use of salt during meal preparation and choose fresh foods over processed ones, which often contain high levels of sodium.
6. **Limit added sugars**: Added sugars can contribute to weight gain, a risk factor for heart disease. It is important to avoid sugary drinks, sweets, and processed snacks, opting instead for healthier alternatives like fresh fruit and natural yogurt.
7. **Stay hydrated**: Water is essential for the proper functioning of the heart and blood vessels. Drinking enough water each day helps maintain blood volume and keep blood pressure under control.
8. **Exercise regularly**: Regular physical activity, such as walking, swimming, or yoga, is crucial for keeping the heart strong and healthy. At least 150 minutes of moderate exercise or 75 minutes of vigorous exercise per week is recommended.
9. **Manage stress**: Chronic stress can contribute to high blood pressure and other heart problems. Stress management techniques such as meditation, deep breathing, and progressive muscle relaxation can help maintain calm and promote heart health.
10. **Don't smoke**: Smoking is one of the main risk factors for heart disease. Quitting smoking can immediately improve cardiovascular health and significantly reduce the risk of heart attacks and strokes.
11. **Limit alcohol consumption**: Excessive alcohol can raise blood pressure and contribute to heart damage. If you choose to drink, it's important to do so in moderation, following guidelines that recommend no more than one drink per day for women and two for men.

Natural Remedies for Weight Management

Zesty Ginger-Lime Salmon with Mango Salsa

Prep Time: 10 minutes
Cook Time: 15-20 minutes
Portion Size: 2

Ingredients:
- 2 salmon fillets (6 oz each)
- 1 tablespoon grated fresh ginger
- Zest and juice of 1 lime
- 1 tablespoon olive oil
- Salt and pepper to taste
- Mango salsa (recipe below)

Mango Salsa:
- 1 ripe mango, diced
- 1/2 red onion, diced
- 1/4 cup chopped cilantro
- 1 tablespoon lime juice
- Pinch of salt

Instructions:
1. Preheat oven to 400°F (200°C).
2. Place salmon fillets on a baking sheet lined with parchment paper.
3. In a small bowl, combine ginger, lime zest and juice, olive oil, salt, and pepper. Rub mixture over salmon.
4. Bake for 15-20 minutes, or until salmon is cooked through.
5. To make mango salsa, combine mango, red onion, cilantro, lime juice, and salt in a bowl.
6. Serve salmon topped with mango salsa.

Nutritional Information (per serving):
- Calories: 350
- Carbohydrates: 20g
- Protein: 30g
- Fat: 15g

Metabolism-Boosting Turkey and Veggie Chili

Prep Time: 15 minutes
Cook Time: 45 minutes
Portion Size: 6

Ingredients:
- 1 tablespoon olive oil
- 1 onion, chopped
- 2 cloves garlic, minced
- 1 pound ground turkey
- 1 red bell pepper, chopped
- 1 green bell pepper, chopped
- 2 stalks celery, chopped
- 1 can (15 ounces) diced tomatoes, undrained
- 1 can (15 ounces) kidney beans, rinsed and drained
- 1 can (15 ounces) black beans, rinsed and drained
- 1 cup vegetable broth
- 1 tablespoon chili powder
- 1 teaspoon cumin
- 1/2 teaspoon paprika
- 1/4 teaspoon cayenne pepper
- Salt and pepper to taste

Instructions:
1. Heat olive oil in a large pot over medium heat. Add onion, garlic, and bell peppers. Cook until softened.
2. Add ground turkey and cook until browned. Drain any excess grease.
3. Add celery, tomatoes, kidney beans, black beans, broth, chili powder, cumin, paprika, cayenne pepper, salt, and pepper.
4. Bring to a boil, then reduce heat and simmer for 30 minutes.
5. Serve with your favorite toppings, such as avocado, chopped cilantro, or Greek yogurt.

Nutritional Information (per serving):
- Calories: 300
- Carbohydrates: 30g
- Protein: 25g
- Fat: 10g

Energizing Quinoa Breakfast Bowl with Berries and Nuts

Prep Time: 5 minutes
Cook Time: 15 minutes
Portion Size: 2

Ingredients:
- 1 cup quinoa, cooked
- 1 cup unsweetened almond milk
- ½ cup mixed berries (blueberries, raspberries, strawberries)
- 2 tablespoons chopped walnuts or pecans

- 1 tablespoon chia seeds
- Drizzle of honey or maple syrup (optional)

Instructions:
1. In a saucepan, combine quinoa and almond milk. Bring to a simmer.
2. Cook for 10-15 minutes, or until quinoa is tender and milk is absorbed.
3. Divide quinoa between two bowls. Top with berries, nuts, and chia seeds.
4. Drizzle with honey or maple syrup, if desired.

Nutritional Information (per serving):
- Calories: 300
- Carbohydrates: 45g
- Protein: 10g
- Fat: 10g

Spicy Black Bean and Sweet Potato Power Bowl

Prep Time: 10 minutes **Cook Time:** 25 minutes **Portion Size:** 2

Ingredients:
- 1 large sweet potato, diced
- 1 can black beans, rinsed and drained
- 1 tablespoon olive oil
- 1/2 teaspoon cumin
- 1/4 teaspoon chili powder
- 1/4 teaspoon smoked paprika
- Pinch of cayenne pepper
- Salt and pepper to taste
- 1 avocado, sliced
- 1/4 cup chopped cilantro
- Lime wedges, for serving

Instructions:
1. Preheat oven to 400°F (200°C).
2. Toss sweet potato with olive oil, cumin, chili powder, paprika, cayenne, salt, and pepper.
3. Roast for 20-25 minutes, or until tender.
4. While sweet potatoes roast, warm black beans in a saucepan.
5. Divide sweet potatoes and black beans between two bowls.
6. Top with avocado slices, cilantro, and a squeeze of lime.

Nutritional Information (per serving):
- Calories: 400
- Carbohydrates: 55g
- Protein: 15g
- Fat: 18g

Revitalizing Green Smoothie with Spinach and Pineapple

Prep Time: 5 minutes
Cook Time: None
Portion Size: 1

Ingredients:
- 1 cup spinach
- 1/2 cup frozen pineapple chunks
- 1/2 banana
- 1/2 cup unsweetened almond milk
- 1 tablespoon chia seeds (optional)

Instructions:
1. Combine all ingredients in a blender.
2. Blend until smooth and creamy.
3. Add more almond milk if needed for desired consistency.

Nutritional Information (per serving):
- Calories: 200
- Carbohydrates: 35g
- Protein: 5g
- Fat: 5g

Aromatic Lentil Soup with Turmeric and Ginger

Prep Time: 10 minutes
Cook Time: 30 minutes
Portion Size: 4

Ingredients:
- 1 tablespoon olive oil
- 1 onion, chopped
- 2 cloves garlic, minced
- 1 teaspoon turmeric
- 1/2 teaspoon ground ginger
- 1 cup red lentils, rinsed
- 4 cups vegetable broth
- Salt and pepper to taste
- 1/4 cup chopped cilantro, for serving

Instructions:
1. Heat olive oil in a large pot over medium heat.
2. Sauté onion until softened.

3. Add garlic, turmeric, and ginger; cook for 1 minute.
4. Stir in lentils and vegetable broth.
5. Bring to a boil, then reduce heat and simmer for 25-30 minutes, or until lentils are tender.
6. Season with salt and pepper.
7. Serve topped with cilantro.

Nutritional Information (per serving):
- Calories: 250
- Carbohydrates: 35g
- Protein: 15g
- Fat: 5g

Colorful Salad with Avocado Dressing

Prep Time: 15 minutes
Cook Time: None
Portion Size: 2

Ingredients:
- 4 cups mixed greens
- 1/2 cup shredded carrots
- 1/2 cup chopped cucumber
- 1/4 cup chopped red onion
- 1/4 cup sunflower seeds

For the dressing:
- 1 avocado
- 1/4 cup plain yogurt
- 2 tablespoons lemon juice
- 1 tablespoon olive oil
- Salt and pepper to taste

Instructions:
1. Combine salad ingredients in a large bowl.
2. In a blender, combine dressing ingredients and blend until smooth.
3. Drizzle dressing over salad and toss to coat.

Nutritional Information (per serving):
- Calories: 350
- Carbohydrates: 20g
- Protein: 10g
- Fat: 25g

Fire-Roasted Chicken with Metabolism-Boosting Spices

Prep Time: 10 minutes
Cook Time: 45 minutes
Portion Size: 4

Ingredients:
- 4 boneless, skinless chicken breasts
- 1 tablespoon olive oil
- 1 teaspoon smoked paprika
- 1/2 teaspoon cayenne pepper
- 1/4 teaspoon garlic powder
- 1/4 teaspoon onion powder
- Salt and pepper to taste
- 1 lemon, sliced

Instructions:
1. Preheat oven to 400°F (200°C).
2. Rub chicken breasts with olive oil and spices.
3. Place chicken and lemon slices on a baking sheet.
4. Bake for 40-45 minutes, or until chicken is cooked through.

Nutritional Information (per serving):
- Calories: 250
- Carbohydrates: 5g
- Protein: 30g
- Fat: 10g

Protein-Packed Egg Muffins with Spinach and Feta

Prep Time: 10 minutes
Cook Time: 20 minutes
Portion Size: 6

Ingredients:
- 6 eggs
- 1/2 cup chopped spinach
- 1/4 cup crumbled feta cheese
- Salt and pepper to taste

Instructions:
1. Preheat oven to 350°F (175°C).
2. Grease a muffin tin.
3. Whisk eggs in a bowl.

4. Stir in spinach and feta.
5. Season with salt and pepper.
6. Divide mixture evenly among muffin cups.
7. Bake for 15-20 minutes, or until set.

Nutritional Information (per serving):
- Calories: 100
- Carbohydrates: 2g
- Protein: 10g
- Fat: 7g

Whimsical Watermelon and Feta Salad with Mint

Prep Time: 10 minutes
Cook Time: None
Portion Size: 4

Ingredients:
- 4 cups cubed watermelon
- 1/2 cup crumbled feta cheese
- 1/4 cup chopped fresh mint
- 2 tablespoons balsamic glaze

Instructions:
1. Combine watermelon, feta, and mint in a large bowl.
2. Drizzle with balsamic glaze and toss gently.

Nutritional Information (per serving):
- Calories: 150
- Carbohydrates: 15g
- Protein: 5g
- Fat: 8g

Sizzling Shrimp Stir-Fry with Zesty Ginger Sauce

Prep Time: 10 minutes
Cook Time: 15 minutes
Portion Size: 2

Ingredients:
- 1 pound shrimp, peeled and deveined
- 1 tablespoon olive oil
- 1/2 cup chopped broccoli florets

- 1/2 cup chopped red bell pepper
- 1/4 cup chopped onion
- 2 cloves garlic, minced
- 1 tablespoon grated ginger
- 2 tablespoons soy sauce
- 1 tablespoon honey
- 1/4 teaspoon red pepper flakes

Instructions:
1. Heat olive oil in a large skillet or wok over medium-high heat.
2. Add shrimp and cook until pink and opaque.
3. Remove shrimp from skillet and set aside.
4. Add broccoli, bell pepper, onion, and garlic to skillet.
5. Stir-fry for 5 minutes, or until vegetables are tender-crisp.
6. Stir in ginger, soy sauce, honey, and red pepper flakes.
7. Return shrimp to skillet and toss to coat.

Nutritional Information (per serving):
- Calories: 300
- Carbohydrates: 15g
- Protein: 30g
- Fat: 12g

Wholesome Turkey Lettuce Wraps with Peanut Sauce

Prep Time: 15 minutes
Cook Time: 10 minutes
Portion Size: 4

Ingredients:
- 1 pound ground turkey
- 1 tablespoon olive oil
- 1/2 cup chopped onion
- 2 cloves garlic, minced
- 1/4 cup chopped water chestnuts
- 1/4 cup hoisin sauce
- 2 tablespoons soy sauce
- 1 tablespoon rice vinegar
- 1 teaspoon sesame oil
- 1/4 teaspoon red pepper flakes
- 1 head butter lettuce, leaves separated

For the peanut sauce:
- 1/4 cup peanut butter

- 2 tablespoons soy sauce
- 1 tablespoon lime juice
- 1 tablespoon honey
- 1 teaspoon grated ginger

Instructions:
1. Heat olive oil in a large skillet over medium heat.
2. Add turkey and cook until browned.
3. Add onion and garlic; cook until softened.
4. Stir in water chestnuts, hoisin sauce, soy sauce, rice vinegar, sesame oil, and red pepper flakes.
5. Simmer for 5 minutes, or until heated through.
6. In a small bowl, whisk together peanut sauce ingredients.
7. To assemble wraps, spoon turkey mixture into lettuce leaves.
8. Top with peanut sauce.

Nutritional Information (per serving):
- Calories: 350
- Carbohydrates: 15g
- Protein: 25g
- Fat: 20g

Vibrant Veggie Omelet with a Kick of Cayenne

Prep Time: 5 minutes
Cook Time: 10 minutes
Portion Size: 1

Ingredients:
- 2 eggs
- 1/4 cup chopped onion
- 1/4 cup chopped bell pepper
- 1/4 cup chopped mushrooms
- 1/4 teaspoon cayenne pepper
- Salt and pepper to taste
- 1 tablespoon olive oil

Instructions:
1. Whisk eggs in a bowl.
2. Stir in onion, bell pepper, mushrooms, and cayenne pepper.
3. Season with salt and pepper.
4. Heat olive oil in a nonstick skillet over medium heat.
5. Pour egg mixture into skillet.
6. Cook for 5-7 minutes, or until set.
7. Fold omelet in half and serve.

Nutritional Information (per serving):
- Calories: 200
- Carbohydrates: 5g
- Protein: 15g
- Fat: 14g

Creamy Avocado and Chicken Salad on Whole-Wheat Bread

Prep Time: 10 minutes
Cook Time: None
Portion Size: 2

Ingredients:
- 2 cups cooked chicken, shredded
- 1 avocado, mashed
- 1/4 cup chopped celery
- 1/4 cup chopped red onion
- 2 tablespoons plain yogurt
- 1 tablespoon lemon juice
- Salt and pepper to taste
- 4 slices whole-wheat bread

Instructions:
1. In a bowl, combine chicken, avocado, celery, onion, yogurt, and lemon juice.
2. Season with salt and pepper.
3. Spread chicken salad on bread slices.

Nutritional Information (per serving):
- Calories: 400
- Carbohydrates: 30g
- Protein: 30g
- Fat: 15g

Spicy Black Bean Burgers with Avocado Crema

Prep Time: 15 minutes
Cook Time: 10 minutes
Portion Size: 4

Ingredients:
- 1 can black beans, rinsed and drained

- 1/2 cup cooked brown rice
- 1/4 cup chopped onion
- 1/4 cup chopped cilantro
- 1 egg, beaten
- 1/4 cup breadcrumbs
- 1 teaspoon chili powder
- 1/2 teaspoon cumin
- Salt and pepper to taste
- 1 avocado
- 1/4 cup plain yogurt
- 1 tablespoon lime juice

Instructions:
1. Mash black beans in a large bowl.
2. Stir in rice, onion, cilantro, egg, breadcrumbs, chili powder, cumin, salt, and pepper.
3. Form mixture into 4 patties.
4. Heat a skillet over medium heat and cook patties for 5 minutes per side, or until browned and heated through.
5. In a small bowl, mash avocado with yogurt and lime juice.
6. Serve burgers topped with avocado crema.

Nutritional Information (per serving):
- Calories: 300
- Carbohydrates: 35g
- Protein: 15g
- Fat: 10g

Refreshing Cucumber and Tomato Salad with Dill Dressing

Prep Time: 10 minutes
Cook Time: None
Portion Size: 4

Ingredients:
- 2 cucumbers, sliced
- 2 cups cherry tomatoes, halved
- 1/4 cup chopped red onion
- 1/4 cup chopped fresh dill
- 2 tablespoons olive oil
- 2 tablespoons lemon juice
- Salt and pepper to taste

Instructions:
1. Combine cucumbers, tomatoes, onion, and dill in a large bowl.
2. Whisk together olive oil, lemon juice, salt, and pepper.

3. Pour dressing over salad and toss to coat.

Nutritional Information (per serving):
- Calories: 100
- Carbohydrates: 10g
- Protein: 2g
- Fat: 7g

Metabolism-Boosting Chicken Soup with Ginger and Turmeric

Prep Time: 10 minutes
Cook Time: 30 minutes
Portion Size: 4

Ingredients:
- 1 tablespoon olive oil
- 1 onion, chopped
- 2 cloves garlic, minced
- 1 teaspoon turmeric
- 1/2 teaspoon ground ginger
- 4 cups chicken broth
- 2 cups cooked chicken, shredded
- 1/2 cup chopped carrots
- 1/4 cup chopped celery
- Salt and pepper to taste
- 1/4 cup chopped cilantro, for serving

Instructions:
1. Heat olive oil in a large pot over medium heat.
2. Sauté onion until softened.
3. Add garlic, turmeric, and ginger; cook for 1 minute.
4. Stir in chicken broth, chicken, carrots, and celery.
5. Bring to a boil, then reduce heat and simmer for 20 minutes.
6. Season with salt and pepper.
7. Serve topped with cilantro.

Nutritional Information (per serving):
- Calories: 250
- Carbohydrates: 10g
- Protein: 25g
- Fat: 10g

Spicy Tuna Salad with a Mediterranean Twist

Prep Time: 10 minutes
Cook Time: None
Portion Size: 2

Ingredients:
- 2 cans tuna, drained
- 1/4 cup chopped Kalamata olives
- 1/4 cup chopped red onion
- 1/4 cup chopped cucumber
- 2 tablespoons chopped fresh parsley
- 2 tablespoons lemon juice
- 1 tablespoon olive oil
- 1/2 teaspoon red pepper flakes
- Salt and pepper to taste

Instructions:
1. In a bowl, combine tuna, olives, onion, cucumber, parsley, lemon juice, olive oil, red pepper flakes, salt, and pepper.
2. Serve on whole-wheat bread or crackers.

Nutritional Information (per serving):
- Calories: 250
- Carbohydrates: 5g
- Protein: 30g
- Fat: 12g

Sweet Potato and Black Bean Stuffed Peppers

Prep Time: 15 minutes
Cook Time: 45 minutes
Portion Size: 4

Ingredients:
- 4 bell peppers, halved and seeded
- 1 tablespoon olive oil
- 1 onion, chopped
- 2 cloves garlic, minced
- 1 sweet potato, diced
- 1 can black beans, rinsed and drained
- 1/2 cup cooked quinoa
- 1/2 cup salsa

- 1/4 cup chopped cilantro
- Salt and pepper to taste

Instructions:
1. Preheat oven to 375°F (190°C).
2. Heat olive oil in a skillet over medium heat.
3. Sauté onion until softened.
4. Add garlic and sweet potato; cook for 5 minutes.
5. Stir in black beans, quinoa, salsa, cilantro, salt, and pepper.
6. Fill pepper halves with mixture.
7. Bake for 30-45 minutes, or until peppers are tender.

Nutritional Information (per serving):
- Calories: 300
- Carbohydrates: 45g
- Protein: 12g
- Fat: 8g

Detoxifying Green Smoothie with Kale, Apple, and Ginger

Prep Time: 5 minutes
Cook Time: None
Portion Size: 1

Ingredients:
- 1 cup kale
- 1/2 apple, cored and chopped
- 1/2 inch piece ginger, peeled and grated
- 1/2 cup unsweetened almond milk
- 1 tablespoon lemon juice

Instructions:
1. Combine all ingredients in a blender.
2. Blend until smooth and creamy.

Nutritional Information (per serving):
- Calories: 150
- Carbohydrates: 25g
- Protein: 5g
- Fat: 5g

Useful tips for weight management

Incorporating herbs and spices into your diet can be a flavorful and healthy way to support your weight loss goals. Herbs like cayenne pepper, ginger, and turmeric have been linked to increased metabolism and reduced appetite. However, it's important to remember that there are no "shortcuts" to weight loss. Herbs and spices are helpful tools, but they should be part of a healthy lifestyle that includes a balanced diet and regular physical activity. The key to long-term success is making sustainable changes that promote overall well-being. Experiment with different herbs and spices to discover which flavors you enjoy most and savor their potential health benefits. Remember, moderation is key: use herbs and spices to enhance the flavor of your dishes, not to mask unhealthy foods.

Here's a list of herbs and spices that may support your weight loss goals:

- **Cayenne pepper:** Boosts metabolism and suppresses appetite.
- **Ginger:** Aids digestion and may reduce inflammation.
- **Turmeric:** Contains curcumin, a potent antioxidant with anti-inflammatory properties.
- **Cinnamon:** May help regulate blood sugar levels and reduce cravings for sweets.
- **Cumin:** Promotes digestion and may help reduce body fat.
- **Black pepper:** Contains piperine, which may increase metabolism and reduce fat absorption.
- **Ginseng:** May boost energy levels and improve exercise endurance.
- **Cardamom:** May help reduce bloating and promote digestion.
- **Mustard seeds:** Contain compounds that may increase metabolism.

Natural Remedies for Blood Sugar Balance

Savory Salmon Cakes with Zesty Dill Yogurt Sauce

Prep Time: 15 minutes
Cook Time: 10 minutes
Portion Size: 2

Ingredients:
- 1 can (6 oz) wild salmon, drained
- 1 egg
- ¼ cup almond flour
- 2 tablespoons chopped green onions
- 1 tablespoon Dijon mustard
- 1 tablespoon lemon juice
- Salt and pepper to taste
- ½ cup plain Greek yogurt
- 1 tablespoon fresh dill, chopped
- 1 teaspoon lemon zest

Instructions:
1. In a bowl, combine salmon, egg, almond flour, green onions, Dijon mustard, lemon juice, salt, and pepper.
2. Form mixture into small patties.
3. Heat a non-stick skillet over medium heat and cook patties for 3-4 minutes on each side until golden brown.
4. In a small bowl, mix Greek yogurt, dill, and lemon zest.
5. Serve salmon cakes with the dill yogurt sauce.

Nutritional Information (per serving):
- Calories: 280
- Carbohydrates: 6g
- Protein: 30g
- Fat: 15g

Harvest Veggie and Lentil Power Bowl with Lemon-Tahini Dressing

Prep Time: 10 minutes
Cook Time: 20 minutes
Portion Size: 2

Ingredients:
- 1 cup cooked lentils
- 1 sweet potato, diced

- 1 cup chopped kale
- ½ cup cherry tomatoes, halved
- 1 avocado, sliced
- 2 tablespoons olive oil
- Salt and pepper to taste
- 2 tablespoons tahini
- 1 tablespoon lemon juice
- 1 teaspoon maple syrup
- 1 garlic clove, minced

Instructions:

1. Preheat oven to 400°F (200°C). Toss sweet potato with 1 tablespoon olive oil, salt, and pepper. Roast for 20 minutes.
2. In a bowl, combine cooked lentils, roasted sweet potato, kale, cherry tomatoes, and avocado.
3. For the dressing, mix tahini, lemon juice, maple syrup, garlic, and water until smooth.
4. Drizzle dressing over the bowl and serve.

Nutritional Information (per serving):
- Calories: 450
- Carbohydrates: 60g
- Protein: 15g
- Fat: 20g

Spiced Chickpea and Sweet Potato Curry with Coconut Milk

Prep Time: 10 minutes
Cook Time: 30 minutes
Portion Size: 2

Ingredients:
- 1 sweet potato, peeled and cubed
- 1 can (15 oz) chickpeas, drained and rinsed
- 1 cup coconut milk
- 1 cup vegetable broth
- 1 onion, chopped
- 2 cloves garlic, minced
- 1 tablespoon curry powder
- 1 teaspoon ground cumin
- 1 teaspoon ground turmeric
- Salt and pepper to taste
- 2 tablespoons olive oil

Instructions:

1. Heat olive oil in a large pot over medium heat. Add onion and cook until translucent.

2. Add garlic, curry powder, cumin, and turmeric, cooking for another minute.
3. Stir in sweet potatoes, chickpeas, coconut milk, and vegetable broth.
4. Bring to a simmer and cook for 20-25 minutes, or until sweet potatoes are tender.
5. Season with salt and pepper to taste and serve hot.

Nutritional Information (per serving):
- Calories: 350
- Carbohydrates: 50g
- Protein: 10g
- Fat: 15g

Hearty Black Bean Soup with Avocado and Lime Crema

Prep Time: 15 minutes
Cook Time: 35 minutes
Portion Size: 2

Ingredients:
- 1 can (15 oz) black beans, drained and rinsed
- 1 onion, chopped
- 2 cloves garlic, minced
- 1 carrot, diced
- 1 celery stalk, diced
- 1 can (14.5 oz) diced tomatoes
- 3 cups vegetable broth
- 1 teaspoon cumin
- 1 teaspoon chili powder
- Salt and pepper to taste
- 1 avocado, sliced
- ¼ cup plain Greek yogurt
- 1 tablespoon lime juice
- 2 tablespoons olive oil

Instructions:
1. Heat olive oil in a large pot over medium heat. Add onion, carrot, and celery, cooking until softened.
2. Add garlic, cumin, and chili powder, cooking for another minute.
3. Stir in black beans, diced tomatoes, and vegetable broth. Bring to a boil.
4. Reduce heat and simmer for 25-30 minutes.
5. In a small bowl, mix Greek yogurt and lime juice to make the crema.
6. Serve soup topped with avocado slices and lime crema.

Nutritional Information (per serving):
- Calories: 300
- Carbohydrates: 45g

- Protein: 12g
- Fat: 10g

Flavorful Chicken Fajita Salad with Avocado-Lime Ranch

Prep Time: 15 minutes
Cook Time: 15 minutes
Portion Size: 2

Ingredients:
- 1 chicken breast, sliced
- 1 bell pepper, sliced
- 1 onion, sliced
- 2 tablespoons olive oil
- 1 teaspoon chili powder
- 1 teaspoon cumin
- 1 teaspoon garlic powder
- Salt and pepper to taste
- 4 cups mixed greens
- 1 avocado, diced
- ¼ cup plain Greek yogurt
- 1 tablespoon lime juice
- 1 tablespoon chopped cilantro

Instructions:
1. Heat olive oil in a skillet over medium heat. Add chicken, bell pepper, and onion. Season with chili powder, cumin, garlic powder, salt, and pepper. Cook until chicken is done and vegetables are tender.
2. In a small bowl, mix Greek yogurt, lime juice, and cilantro to make the avocado-lime ranch dressing.
3. In a large bowl, combine mixed greens, avocado, and cooked chicken and vegetables.
4. Drizzle with avocado-lime ranch dressing and serve.

Nutritional Information (per serving):
- Calories: 350
- Carbohydrates: 20g
- Protein: 30g
- Fat: 18g

Spicy Shrimp and Vegetable Stir-Fry with Brown Rice

Prep Time: 10 minutes

Cook Time: 15 minutes
Portion Size: 2

Ingredients:
- 8 oz shrimp, peeled and deveined
- 1 cup broccoli florets
- 1 bell pepper, sliced
- 1 carrot, sliced
- 2 tablespoons soy sauce
- 1 tablespoon sriracha sauce
- 2 tablespoons olive oil
- 1 teaspoon ginger, minced
- 2 cloves garlic, minced
- 1 cup cooked brown rice

Instructions:
1. Heat olive oil in a wok or large skillet over medium-high heat. Add ginger and garlic, cooking until fragrant.
2. Add shrimp and cook until pink. Remove shrimp and set aside.
3. Add broccoli, bell pepper, and carrot to the wok, cooking until tender-crisp.
4. Stir in soy sauce and sriracha sauce.
5. Return shrimp to the wok and toss to coat.
6. Serve stir-fry over cooked brown rice.

Nutritional Information (per serving):
- Calories: 400
- Carbohydrates: 40g
- Protein: 30g
- Fat: 15g

Turkey Meatballs with Zucchini Noodles and Marinara Sauce

Prep Time: 15 minutes
Cook Time: 20 minutes
Portion Size: 2

Ingredients:
- 8 oz ground turkey
- 1 egg
- ¼ cup almond flour
- 1 teaspoon garlic powder
- 1 teaspoon onion powder
- Salt and pepper to taste
- 2 zucchinis, spiralized

- 1 cup marinara sauce
- 2 tablespoons olive oil

Instructions:
1. In a bowl, combine ground turkey, egg, almond flour, garlic powder, onion powder, salt, and pepper. Form into meatballs.
2. Heat olive oil in a skillet over medium heat. Add meatballs and cook until browned and cooked through.
3. In a separate skillet, heat marinara sauce.
4. Add zucchini noodles to the marinara sauce and cook until tender.
5. Serve meatballs over zucchini noodles with marinara sauce.

Nutritional Information (per serving):
- Calories: 350
- Carbohydrates: 15g
- Protein: 30g
- Fat: 20g

Baked Cod with Roasted Brussels Sprouts and Lemon-Garlic Drizzle

Prep Time: 10 minutes
Cook Time: 20 minutes
Portion Size: 2

Ingredients:
- 2 cod fillets (4 oz each)
- 2 cups Brussels sprouts, halved
- 3 tablespoons olive oil
- 1 lemon, zested and juiced
- 2 cloves garlic, minced
- Salt and pepper to taste

Instructions:
1. Preheat oven to 400°F (200°C). Toss Brussels sprouts with 1 tablespoon olive oil, salt, and pepper. Roast for 15 minutes.
2. Place cod fillets on a baking sheet. Drizzle with 1 tablespoon olive oil, salt, and pepper. Bake for 10-12 minutes.
3. In a small bowl, mix remaining olive oil, lemon zest, lemon juice, and garlic.
4. Serve cod and Brussels sprouts drizzled with lemon-garlic mixture.

Nutritional Information (per serving):
- Calories: 280
- Carbohydrates: 15g
- Protein: 25g

- Fat: 15g

Spinach and Feta Stuffed Chicken Breast with Balsamic Glaze

Prep Time: 15 minutes
Cook Time: 25 minutes
Portion Size: 2

Ingredients:
- 2 chicken breasts
- 1 cup fresh spinach, chopped
- ½ cup crumbled feta cheese
- 2 tablespoons olive oil
- ¼ cup balsamic vinegar
- 1 tablespoon honey
- Salt and pepper to taste

Instructions:
1. Preheat oven to 375°F (190°C).
2. Slice chicken breasts horizontally to create pockets.
3. Stuff each pocket with spinach and feta cheese, then secure with toothpicks.
4. Heat olive oil in a skillet over medium-high heat. Sear chicken breasts on both sides until golden brown.
5. Transfer chicken to a baking dish and bake for 15-20 minutes until cooked through.
6. In the same skillet, add balsamic vinegar and honey, cooking until reduced and thickened.
7. Drizzle balsamic glaze over chicken before serving.

Nutritional Information (per serving):
- Calories: 320
- Carbohydrates: 12g
- Protein: 35g
- Fat: 15g

Slow-Cooker Beef Stew with Root Vegetables and Herbs

Prep Time: 15 minutes
Cook Time: 8 hours
Portion Size: 2

Ingredients:

- 8 oz beef stew meat
- 1 potato, diced
- 1 carrot, sliced
- 1 celery stalk, sliced
- 1 onion, chopped
- 2 cups beef broth
- 1 teaspoon dried thyme
- 1 teaspoon dried rosemary
- Salt and pepper to taste

Instructions:
1. Place beef, potato, carrot, celery, and onion in a slow cooker.
2. Add beef broth, thyme, rosemary, salt, and pepper.
3. Cook on low for 8 hours or until beef and vegetables are tender.
4. Serve hot.

Nutritional Information (per serving):
- Calories: 350
- Carbohydrates: 30g
- Protein: 25g
- Fat: 15g

One-Pan Roasted Chicken with Rainbow Vegetables and Herbs de Provence

Prep Time: 10 minutes
Cook Time: 35 minutes
Portion Size: 2

Ingredients:
- 2 chicken thighs
- 1 red bell pepper, sliced
- 1 yellow bell pepper, sliced
- 1 zucchini, sliced
- 1 red onion, sliced
- 2 tablespoons olive oil
- 1 teaspoon Herbs de Provence
- Salt and pepper to taste

Instructions:
1. Preheat oven to 425°F (220°C).
2. Toss bell peppers, zucchini, and red onion with 1 tablespoon olive oil, Herbs de Provence, salt, and pepper. Spread on a baking sheet.

3. Rub chicken thighs with remaining olive oil, salt, and pepper. Place on top of vegetables.
4. Roast for 35 minutes or until chicken is cooked through and vegetables are tender.

Nutritional Information (per serving):
- Calories: 400
- Carbohydrates: 20g
- Protein: 30g
- Fat: 25g

Mediterranean Quinoa Salad with Chickpeas, Cucumber, and Feta

Prep Time: 10 minutes
Cook Time: 15 minutes
Portion Size: 2

Ingredients:
- 1 cup cooked quinoa
- 1 can (15 oz) chickpeas, drained and rinsed
- 1 cucumber, diced
- ½ cup cherry tomatoes, halved
- ¼ cup crumbled feta cheese
- 2 tablespoons olive oil
- 1 tablespoon lemon juice
- 1 teaspoon dried oregano
- Salt and pepper to taste

Instructions:
1. In a large bowl, combine cooked quinoa, chickpeas, cucumber, cherry tomatoes, and feta cheese.
2. In a small bowl, whisk together olive oil, lemon juice, oregano, salt, and pepper.
3. Pour dressing over salad and toss to combine.
4. Serve chilled or at room temperature.

Nutritional Information (per serving):
- Calories: 350
- Carbohydrates: 45g
- Protein: 12g
- Fat: 15g

Egg White Frittata with Spinach, Mushrooms, and Goat Cheese

Prep Time: 10 minutes
Cook Time: 15 minutes
Portion Size: 2

Ingredients:
- 8 egg whites
- 1 cup fresh spinach, chopped
- ½ cup mushrooms, sliced
- ¼ cup crumbled goat cheese
- 1 tablespoon olive oil
- Salt and pepper to taste

Instructions:
1. Preheat oven to 375°F (190°C).
2. Heat olive oil in an oven-safe skillet over medium heat. Add mushrooms and spinach, cooking until tender.
3. Pour egg whites over vegetables and cook until edges start to set.
4. Sprinkle goat cheese on top and transfer skillet to the oven.
5. Bake for 10-12 minutes or until eggs are fully set.
6. Serve hot.

Nutritional Information (per serving):
- Calories: 200
- Carbohydrates: 5g
- Protein: 20g
- Fat: 10g

Chia Seed Pudding with Mixed Berries and Almonds

Prep Time: 10 minutes (plus overnight chilling)
Cook Time: 0 minutes
Portion Size: 2

Ingredients:
- ¼ cup chia seeds
- 1 cup almond milk
- 1 tablespoon honey or maple syrup
- ½ cup mixed berries (blueberries, raspberries, strawberries)
- 2 tablespoons sliced almonds

Instructions:
1. In a bowl, mix chia seeds, almond milk, and honey or maple syrup.
2. Stir well and refrigerate overnight or for at least 4 hours.
3. Stir again before serving and top with mixed berries and sliced almonds.

Nutritional Information (per serving):
- Calories: 250
- Carbohydrates: 25g
- Protein: 6g
- Fat: 15g

Avocado Toast with Smoked Salmon and Everything Bagel Seasoning

Prep Time: 5 minutes
Cook Time: 5 minutes
Portion Size: 2

Ingredients:
- 2 slices whole grain bread
- 1 avocado, mashed
- 2 oz smoked salmon
- 1 teaspoon Everything Bagel seasoning

Instructions:
1. Toast the bread slices to your desired level of crispiness.
2. Spread mashed avocado evenly on each toast slice.
3. Top with smoked salmon and sprinkle with Everything Bagel seasoning.

Nutritional Information (per serving):
- Calories: 300
- Carbohydrates: 30g
- Protein: 12g
- Fat: 18g

Spicy Black Bean Burgers with Mango Salsa and Guacamole

Prep Time: 20 minutes
Cook Time: 10 minutes
Portion Size: 2

Ingredients:
- 1 can (15 oz) black beans, drained and rinsed
- ¼ cup breadcrumbs
- 1 egg
- 1 teaspoon chili powder

- 1 teaspoon cumin
- Salt and pepper to taste
- 1 mango, diced
- 1 tablespoon red onion, minced
- 1 tablespoon cilantro, chopped
- 1 avocado, mashed
- 2 whole grain burger buns

Instructions:
1. In a bowl, mash black beans until mostly smooth. Add breadcrumbs, egg, chili powder, cumin, salt, and pepper. Mix well and form into patties.
2. Heat a skillet over medium heat and cook patties for 5 minutes on each side until crispy.
3. In a small bowl, combine mango, red onion, and cilantro to make the salsa.
4. Spread mashed avocado on the bottom half of each bun. Top with black bean patty and mango salsa. Cover with the top bun and serve.

Nutritional Information (per serving):
- Calories: 400
- Carbohydrates: 55g
- Protein: 15g
- Fat: 15g

Turkey Chili with Butternut Squash and Kidney Beans

Prep Time: 15 minutes
Cook Time: 40 minutes
Portion Size: 2

Ingredients:
- 8 oz ground turkey
- 1 cup butternut squash, cubed
- 1 can (15 oz) kidney beans, drained and rinsed
- 1 can (14.5 oz) diced tomatoes
- 1 onion, chopped
- 2 cloves garlic, minced
- 2 tablespoons chili powder
- 1 teaspoon cumin
- Salt and pepper to taste
- 2 tablespoons olive oil

Instructions:
1. Heat olive oil in a large pot over medium heat. Add onion and cook until translucent.
2. Add garlic and ground turkey, cooking until browned.
3. Stir in butternut squash, kidney beans, diced tomatoes, chili powder, cumin, salt, and pepper.

4. Bring to a boil, then reduce heat and simmer for 30 minutes or until butternut squash is tender.
5. Serve hot.

Nutritional Information (per serving):
- Calories: 350
- Carbohydrates: 40g
- Protein: 30g
- Fat: 12g

Greek Yogurt Parfait with Berries, Granola, and Honey

Prep Time: 5 minutes
Cook Time: 0 minutes
Portion Size: 2

Ingredients:
- 1 cup plain Greek yogurt
- ½ cup mixed berries (blueberries, raspberries, strawberries)
- ¼ cup granola
- 2 tablespoons honey

Instructions:
1. In two bowls or glasses, layer Greek yogurt, mixed berries, and granola.
2. Drizzle honey on top and serve immediately.

Calories Chicken Breast with Balsamic Glaze

Prep Time: 15 minutes
Cook Time: 25 minutes
Portion Size: 2

Ingredients:
- 2 chicken breasts
- 1 cup fresh spinach, chopped
- ½ cup crumbled feta cheese
- 2 tablespoons olive oil
- ¼ cup balsamic vinegar
- 1 tablespoon honey
- Salt and pepper to taste

Instructions:
1. Preheat oven to 375°F (190°C).

2. Slice chicken breasts horizontally to create pockets.
3. Stuff each pocket with spinach and feta cheese, then secure with toothpicks.
4. Heat olive oil in a skillet over medium-high heat. Sear chicken breasts on both sides until golden brown.
5. Transfer chicken to a baking dish and bake for 15-20 minutes until cooked through.
6. In the same skillet, add balsamic vinegar and honey, cooking until reduced and thickened.
7. Drizzle balsamic glaze over chicken before serving.

Nutritional Information (per serving):
- Calories: 320
- Carbohydrates: 12g
- Protein: 35g
- Fat: 15g

Cauliflower Rice Stir-Fry with Chicken and Peanut Sauce

Prep Time: 10 minutes
Cook Time: 15 minutes
Portion Size: 2

Ingredients:
- 1 chicken breast, sliced
- 2 cups cauliflower rice
- 1 bell pepper, sliced
- 1 carrot, sliced
- 2 tablespoons peanut butter
- 2 tablespoons soy sauce
- 1 tablespoon lime juice
- 1 tablespoon olive oil
- 1 teaspoon ginger, minced
- 1 clove garlic, minced

Instructions:
1. Heat olive oil in a large skillet over medium-high heat. Add ginger and garlic, cooking until fragrant.
2. Add chicken and cook until browned and cooked through. Remove from skillet and set aside.
3. Add bell pepper and carrot to the skillet, cooking until tender.
4. Stir in cauliflower rice and cook for another 3-4 minutes.
5. In a small bowl, mix peanut butter, soy sauce, and lime juice until smooth.
6. Return chicken to the skillet and pour peanut sauce over the mixture. Toss to coat and serve hot.

Nutritional Information (per serving):
- Calories: 350
- Carbohydrates: 20g

- Protein: 30g
- Fat: 15g

Lemony Lentil Soup with Spinach and Turmeric

Prep Time: 10 minutes
Cook Time: 30 minutes
Portion Size: 2

Ingredients:
- 1 cup lentils, rinsed
- 1 onion, chopped
- 2 cloves garlic, minced
- 1 carrot, diced
- 1 celery stalk, diced
- 4 cups vegetable broth
- 1 teaspoon ground turmeric
- 1 tablespoon lemon juice
- 2 cups fresh spinach, chopped
- 2 tablespoons olive oil
- Salt and pepper to taste

Instructions:
1. Heat olive oil in a large pot over medium heat. Add onion, carrot, and celery, cooking until softened.
2. Add garlic and turmeric, cooking for another minute.
3. Stir in lentils and vegetable broth. Bring to a boil, then reduce heat and simmer for 25 minutes or until lentils are tender.
4. Stir in lemon juice and spinach, cooking until spinach is wilted.
5. Season with salt and pepper to taste and serve hot.

Nutritional Information (per serving):
- Calories: 300
- Carbohydrates: 45g
- Protein: 15g
- Fat: 10g

Useful tips for Blood Sugar Balance

To prevent diabetes, it is crucial to adopt a holistic approach that incorporates diet, lifestyle, and regular monitoring of health markers. Dr. Barbara O'Neill emphasizes the importance of consuming a diet rich in whole, unprocessed foods. Focus on incorporating plenty of fresh vegetables, lean proteins, and healthy fats into your meals, while reducing the intake of refined sugars and carbohydrates. Regular physical activity is also essential, as it helps to maintain a healthy weight and improves insulin sensitivity. Aim for at least 30 minutes of moderate exercise most days of the week. Additionally, managing stress through practices such as meditation, deep breathing, and adequate sleep is vital for overall health and can help regulate blood sugar levels. Regular check-ups with your healthcare provider to monitor blood glucose levels and other risk factors are important to catch any early signs of diabetes and take proactive measures. By following these guidelines, you can significantly reduce your risk of developing diabetes and promote a healthier, more balanced lifestyle.

Natural Remedies for Joint Health

Turmeric-Ginger Anti-Inflammatory Smoothie

Prep Time: 5 minutes
Cook Time: 0 minutes
Portion Size: 2

Ingredients:
- 1 cup unsweetened almond milk
- 1 banana
- 1 teaspoon fresh ginger, grated
- 1 teaspoon ground turmeric
- 1 tablespoon honey
- ½ teaspoon cinnamon
- 1 cup ice cubes

Instructions:
1. Combine all ingredients in a blender.
2. Blend until smooth.
3. Serve immediately.

Nutritional Information (per serving):
- Calories: 150
- Carbohydrates: 30g
- Protein: 2g
- Fat: 3g

Omega-3 Rich Salmon and Avocado Salad

Prep Time: 10 minutes
Cook Time: 10 minutes
Portion Size: 2

Ingredients:
- 2 salmon fillets (4 oz each)
- 1 tablespoon olive oil
- 4 cups mixed greens
- 1 avocado, sliced
- ½ red onion, thinly sliced
- 1 tablespoon lemon juice
- Salt and pepper to taste

Instructions:

1. Heat olive oil in a skillet over medium heat. Cook salmon fillets for 4-5 minutes on each side until fully cooked.
2. In a large bowl, combine mixed greens, avocado, and red onion.
3. Top with cooked salmon and drizzle with lemon juice. Season with salt and pepper.
4. Serve immediately.

Nutritional Information (per serving):
- Calories: 400
- Carbohydrates: 10g
- Protein: 30g
- Fat: 25g

Bone Broth with Fresh Herbs and Garlic

Prep Time: 10 minutes
Cook Time: 10 hours
Portion Size: 2

Ingredients:
- 2 lbs beef bones
- 2 carrots, chopped
- 2 celery stalks, chopped
- 1 onion, quartered
- 4 cloves garlic, smashed
- 1 tablespoon apple cider vinegar
- 8 cups water
- 1 bunch fresh parsley
- Salt and pepper to taste

Instructions:
1. Place bones, carrots, celery, onion, and garlic in a slow cooker. Add apple cider vinegar and water.
2. Cook on low for 10 hours.
3. Strain broth and season with salt and pepper. Garnish with fresh parsley before serving.

Nutritional Information (per serving):
- Calories: 50
- Carbohydrates: 5g
- Protein: 5g
- Fat: 2g

Anti-Inflammatory Sweet Potato and Quinoa Bowl

Prep Time: 10 minutes
Cook Time: 20 minutes
Portion Size: 2

Ingredients:
- 1 cup cooked quinoa
- 1 sweet potato, peeled and cubed
- 1 cup broccoli florets
- 1 avocado, sliced
- 2 tablespoons olive oil
- 1 teaspoon ground turmeric
- Salt and pepper to taste

Instructions:
1. Preheat oven to 400°F (200°C). Toss sweet potato and broccoli with olive oil, turmeric, salt, and pepper. Roast for 20 minutes.
2. In two bowls, divide quinoa, roasted sweet potato, and broccoli.
3. Top with sliced avocado and serve.

Nutritional Information (per serving):
- Calories: 400
- Carbohydrates: 55g
- Protein: 10g
- Fat: 18g

Curcumin-Infused Golden Milk Latte

Prep Time: 5 minutes
Cook Time: 5 minutes
Portion Size: 2

Ingredients:
- 2 cups unsweetened almond milk
- 1 teaspoon ground turmeric
- ½ teaspoon ground ginger
- 1 tablespoon honey
- ¼ teaspoon cinnamon
- Pinch of black pepper

Instructions:
1. In a saucepan, combine almond milk, turmeric, ginger, honey, cinnamon, and black pepper.

2. Heat over medium heat until warm, stirring frequently.
3. Pour into mugs and serve.

Nutritional Information (per serving):
- Calories: 100
- Carbohydrates: 20g
- Protein: 1g
- Fat: 2g

Green Tea and Berry Antioxidant Smoothie

Prep Time: 5 minutes
Cook Time: 0 minutes
Portion Size: 2

Ingredients:
- 1 cup brewed green tea, cooled
- 1 cup mixed berries (blueberries, raspberries, strawberries)
- 1 banana
- ½ cup plain Greek yogurt
- 1 tablespoon honey
- 1 cup ice cubes

Instructions:
1. Combine all ingredients in a blender.
2. Blend until smooth.
3. Serve immediately.

Nutritional Information (per serving):
- Calories: 180
- Carbohydrates: 35g
- Protein: 6g
- Fat: 2g

Collagen-Boosting Chicken Soup with Vegetables

Prep Time: 15 minutes
Cook Time: 45 minutes
Portion Size: 2

Ingredients:

- 1 chicken breast
- 2 carrots, sliced
- 2 celery stalks, sliced
- 1 onion, chopped
- 3 cloves garlic, minced
- 6 cups chicken broth
- 1 teaspoon dried thyme
- 1 teaspoon dried rosemary
- Salt and pepper to taste

Instructions:

1. In a large pot, combine chicken breast, carrots, celery, onion, garlic, chicken broth, thyme, rosemary, salt, and pepper.
2. Bring to a boil, then reduce heat and simmer for 45 minutes.
3. Remove chicken breast, shred it, and return to the pot. Serve hot.

Nutritional Information (per serving):
- Calories: 250
- Carbohydrates: 20g
- Protein: 30g
- Fat: 8g

Anti-Inflammatory Turmeric and Coconut Lentil Stew

Prep Time: 10 minutes
Cook Time: 30 minutes
Portion Size: 2

Ingredients:
- 1 cup red lentils, rinsed
- 1 can (14 oz) coconut milk
- 1 cup vegetable broth
- 1 onion, chopped
- 2 cloves garlic, minced
- 1 tablespoon ground turmeric
- 1 teaspoon ground cumin
- Salt and pepper to taste
- 2 tablespoons olive oil

Instructions:

1. Heat olive oil in a large pot over medium heat. Add onion and cook until translucent.
2. Add garlic, turmeric, and cumin, cooking for another minute.
3. Stir in lentils, coconut milk, and vegetable broth. Bring to a boil, then reduce heat and simmer for 25 minutes or until lentils are tender.

4. Season with salt and pepper and serve.

Nutritional Information (per serving):
- Calories: 350
- Carbohydrates: 45g
- Protein: 15g
- Fat: 15g

Pineapple and Chia Seed Anti-Inflammatory Parfait

Prep Time: 5 minutes
Cook Time: 0 minutes
Portion Size: 2

Ingredients:
- 1 cup pineapple, diced
- 2 tablespoons chia seeds
- 1 cup coconut yogurt
- 1 tablespoon honey
- ¼ cup granola

Instructions:
1. In two bowls or glasses, layer coconut yogurt, pineapple, chia seeds, and granola.
2. Drizzle honey on top and serve immediately.

Nutritional Information (per serving):
- Calories: 250
- Carbohydrates: 40g
- Protein: 6g
- Fat: 8g

Spicy Ginger and Lemongrass Shrimp Stir-Fry

Prep Time: 10 minutes
Cook Time: 10 minutes
Portion Size: 2

Ingredients:
- 8 oz shrimp, peeled and deveined
- 1 cup bell pepper, sliced
- 1 cup snap peas

- 1 stalk lemongrass, minced
- 1 tablespoon fresh ginger, grated
- 2 tablespoons soy sauce
- 1 tablespoon sriracha sauce
- 2 tablespoons olive oil

Instructions:
1. Heat olive oil in a large skillet over medium-high heat. Add lemongrass and ginger, cooking until fragrant.
2. Add shrimp and cook until pink. Remove shrimp and set aside.
3. Add bell pepper and snap peas to the skillet, cooking until tender-crisp.
4. Stir in soy sauce and sriracha sauce.
5. Return shrimp to the skillet and toss to coat. Serve hot.

Nutritional Information (per serving):
- Calories: 300
- Carbohydrates: 15g
- Protein: 25g
- Fat: 15g

Omega-3 Packed Walnut and Flaxseed Porridge

Prep Time: 5 minutes
Cook Time: 10 minutes
Portion Size: 2

Ingredients:
- 1 cup rolled oats
- 2 cups almond milk
- 2 tablespoons ground flaxseed
- ¼ cup walnuts, chopped
- 1 tablespoon honey
- 1 teaspoon cinnamon

Instructions:
1. In a saucepan, combine rolled oats and almond milk. Bring to a simmer and cook for 5-7 minutes until thickened.
2. Stir in ground flaxseed, walnuts, honey, and cinnamon. Serve hot.

Nutritional Information (per serving):
- Calories: 350
- Carbohydrates: 45g
- Protein: 10g
- Fat: 15g

Tart Cherry and Kale Anti-Inflammatory Salad

Prep Time: 10 minutes
Cook Time: 0 minutes
Portion Size: 2

Ingredients:
- 4 cups kale, chopped
- 1 cup tart cherries, pitted and halved
- ¼ cup sliced almonds
- 2 tablespoons olive oil
- 1 tablespoon apple cider vinegar
- 1 tablespoon honey
- Salt and pepper to taste

Instructions:
1. In a large bowl, combine kale, tart cherries, and sliced almonds.
2. In a small bowl, whisk together olive oil, apple cider vinegar, honey, salt, and pepper.
3. Pour dressing over salad and toss to combine. Serve immediately.

Nutritional Information (per serving):
- Calories: 200
- Carbohydrates: 25g
- Protein: 5g
- Fat: 12g

Zesty Lemon and Garlic Broccoli Stir-Fry

Prep Time: 5 minutes
Cook Time: 10 minutes
Portion Size: 2

Ingredients:
- 2 cups broccoli florets
- 2 cloves garlic, minced
- 1 tablespoon olive oil
- 1 tablespoon lemon juice
- 1 teaspoon lemon zest
- Salt and pepper to taste

Instructions:
1. Heat olive oil in a large skillet over medium heat. Add garlic and cook until fragrant.
2. Add broccoli florets and cook until tender-crisp.
3. Stir in lemon juice, lemon zest, salt, and pepper. Serve hot.

Nutritional Information (per serving):
- Calories: 100
- Carbohydrates: 10g
- Protein: 4g
- Fat: 7g

Green Smoothie with Spinach, Avocado, and Flaxseeds

Prep Time: 5 minutes
Cook Time: 0 minutes
Portion Size: 2

Ingredients:
- 1 cup spinach
- 1 avocado
- 1 banana
- 1 tablespoon ground flaxseed
- 1 cup unsweetened almond milk
- 1 cup ice cubes

Instructions:
1. Combine all ingredients in a blender.
2. Blend until smooth.
3. Serve immediately.

Nutritional Information (per serving):
- Calories: 250
- Carbohydrates: 25g
- Protein: 3g
- Fat: 15g

Anti-Inflammatory Baked Turmeric Cauliflower

Prep Time: 10 minutes
Cook Time: 25 minutes

Portion Size: 2

Ingredients:
- 1 head cauliflower, cut into florets
- 2 tablespoons olive oil
- 1 teaspoon ground turmeric
- 1 teaspoon garlic powder
- Salt and pepper to taste

Instructions:

1. Preheat oven to 400°F (200°C). Toss cauliflower florets with olive oil, turmeric, garlic powder, salt, and pepper.
2. Spread on a baking sheet and roast for 25 minutes until tender and golden brown. Serve hot.

Nutritional Information (per serving):
- Calories: 150
- Carbohydrates: 12g
- Protein: 4g
- Fat: 10g

Ginger and Miso Soup with Seaweed

Prep Time: 5 minutes
Cook Time: 10 minutes
Portion Size: 2

Ingredients:
- 4 cups vegetable broth
- 1 tablespoon miso paste
- 1 tablespoon fresh ginger, grated
- 1 cup seaweed, rehydrated
- 1 green onion, sliced

Instructions:

1. In a pot, bring vegetable broth to a simmer. Stir in miso paste and ginger.
2. Add seaweed and cook for 5 minutes.
3. Garnish with sliced green onion and serve hot.

Nutritional Information (per serving):
- Calories: 50
- Carbohydrates: 5g
- Protein: 2g
- Fat: 1g

Spicy Turmeric and Lentil Stew

Prep Time: 10 minutes
Cook Time: 30 minutes
Portion Size: 2

Ingredients:
- 1 cup red lentils, rinsed
- 1 onion, chopped
- 2 cloves garlic, minced
- 1 tablespoon ground turmeric
- 1 teaspoon ground cumin
- 1 teaspoon chili powder
- 4 cups vegetable broth
- 2 tablespoons olive oil
- Salt and pepper to taste

Instructions:
1. Heat olive oil in a pot over medium heat. Add onion and cook until translucent.
2. Add garlic, turmeric, cumin, and chili powder, cooking for another minute.
3. Stir in lentils and vegetable broth. Bring to a boil, then reduce heat and simmer for 25 minutes until lentils are tender.
4. Season with salt and pepper and serve hot.

Nutritional Information (per serving):
- Calories: 300
- Carbohydrates: 45g
- Protein: 15g
- Fat: 10g

Garlic-Roasted Brussels Sprouts with Walnuts

Prep Time: 5 minutes
Cook Time: 20 minutes
Portion Size: 2

Ingredients:
- 2 cups Brussels sprouts, halved
- 2 tablespoons olive oil
- 2 cloves garlic, minced
- ¼ cup walnuts, chopped
- Salt and pepper to taste

Instructions:
1. Preheat oven to 400°F (200°C). Toss Brussels sprouts with olive oil, garlic, salt, and pepper. Spread on a baking sheet and roast for 20 minutes.
2. Sprinkle with walnuts before serving.

Nutritional Information (per serving):
- Calories: 200
- Carbohydrates: 18g
- Protein: 5g
- Fat: 14g

Cucumber and Turmeric Detox Water

Prep Time: 5 minutes
Cook Time: 0 minutes
Portion Size: 2

Ingredients:
- 4 cups water
- 1 cucumber, sliced
- 1 teaspoon ground turmeric
- 1 lemon, sliced
- 1 tablespoon honey

Instructions:
1. In a pitcher, combine water, cucumber, turmeric, lemon, and honey. Stir well.
2. Refrigerate for at least 1 hour before serving.

Nutritional Information (per serving):
- Calories: 20
- Carbohydrates: 5g
- Protein: 0g
- Fat: 0g

Anti-Inflammatory Blueberry and Almond Oatmeal

Prep Time: 5 minutes
Cook Time: 10 minutes
Portion Size: 2

Ingredients:

- 1 cup rolled oats
- 2 cups almond milk
- 1 cup blueberries
- ¼ cup almonds, chopped
- 1 tablespoon honey
- 1 teaspoon ground turmeric

Instructions:

1. In a saucepan, combine rolled oats and almond milk. Bring to a simmer and cook for 5-7 minutes until thickened.
2. Stir in blueberries, almonds, honey, and turmeric. Serve hot.

Nutritional Information (per serving):

- Calories: 300
- Carbohydrates: 50g
- Protein: 8g
- Fat: 12g

Main Anti-Inflammatory and Pain-Relieving Herbs

Turmeric
- **Active Component:** Curcumin
- **Benefits:** Reduces inflammation, pain relief, antioxidant properties
- **Usage:** Can be added to foods, taken as a supplement, or brewed as tea

Ginger
- **Active Component:** Gingerol
- **Benefits:** Anti-inflammatory, pain relief, digestive aid
- **Usage:** Fresh, dried, powdered, or as an extract in teas and foods

Boswellia (Frankincense)
- **Active Component:** Boswellic acids
- **Benefits:** Reduces inflammation, improves joint health, pain relief
- **Usage:** Supplements, essential oils for topical use

Willow Bark
- **Active Component:** Salicin
- **Benefits:** Pain relief, anti-inflammatory, similar to aspirin
- **Usage:** Dried bark used in teas or as a supplement

Devil's Claw
- **Active Component:** Harpagoside
- **Benefits:** Pain relief, reduces inflammation, used for arthritis
- **Usage:** Supplements, dried roots in teas

Feverfew
- **Active Component:** Parthenolide
- **Benefits:** Reduces inflammation, pain relief, used for migraines
- **Usage:** Fresh leaves, dried leaves, or as a supplement

Capsaicin
- **Active Component:** Capsaicinoids
- **Benefits:** Pain relief, reduces inflammation, used in topical creams
- **Usage:** Topical ointments and creams

Green Tea
- **Active Component:** Epigallocatechin gallate (EGCG)
- **Benefits:** Anti-inflammatory, antioxidant, supports overall health
- **Usage:** Brewed as tea, or taken as an extract

Nutritional Strategies for Optimal Eye Health

Zesty Kale and Quinoa Salad with Lemon-Tahini Dressing for Eye-Brightening Vitamin A

Prep Time: 15 minutes
Cook Time: 20 minutes
Portion Size: 2

Ingredients:
- 1 cup cooked quinoa
- 2 cups kale, chopped
- 1 carrot, shredded
- 1/4 cup red bell pepper, diced
- 1/4 cup cucumber, diced
- 2 tablespoons sunflower seeds
- 1/4 cup lemon juice
- 2 tablespoons tahini
- 1 tablespoon olive oil
- 1 clove garlic, minced
- Salt and pepper to taste

Instructions:
1. Cook quinoa according to package instructions and let it cool.
2. In a large bowl, combine kale, carrot, bell pepper, cucumber, and sunflower seeds.
3. In a small bowl, whisk together lemon juice, tahini, olive oil, garlic, salt, and pepper.
4. Pour dressing over the salad and toss to coat.
5. Serve chilled.

Nutritional Information (per serving):
- Calories: 350
- Carbohydrates: 45g
- Protein: 12g
- Fat: 15g

Salmon Power Bowl with Sweet Potato and Spinach for Omega-3 Rich Vision Support

Prep Time: 15 minutes
Cook Time: 25 minutes
Portion Size: 2

Ingredients:

- 2 salmon fillets
- 1 large sweet potato, diced
- 2 cups spinach, chopped
- 1/4 cup red onion, sliced
- 1 avocado, sliced
- 1 tablespoon olive oil
- 1 lemon, juiced
- Salt and pepper to taste

Instructions:

1. Preheat oven to 400°F (200°C).
2. Toss sweet potato with olive oil, salt, and pepper, then spread on a baking sheet.
3. Bake sweet potato for 20-25 minutes until tender.
4. Meanwhile, season salmon with salt and pepper, then bake for 15-20 minutes until cooked through.
5. In a bowl, combine spinach, red onion, and avocado.
6. Add cooked sweet potato and salmon.
7. Drizzle with lemon juice and toss gently.

Nutritional Information (per serving):

- Calories: 450
- Carbohydrates: 30g
- Protein: 35g
- Fat: 20g

Sunshine Smoothie with Mango, Carrot, and Orange for Antioxidant-Packed Eye Protection

Prep Time: 10 minutes
Cook Time: 0 minutes
Portion Size: 2

Ingredients:

- 1 mango, peeled and diced
- 2 carrots, peeled and chopped
- 1 orange, peeled and segmented
- 1 cup coconut water
- 1 tablespoon chia seeds
- 1/2 cup ice

Instructions:

1. Combine mango, carrots, orange, coconut water, chia seeds, and ice in a blender.
2. Blend until smooth.
3. Serve immediately.

Nutritional Information (per serving):
- Calories: 200
- Carbohydrates: 45g
- Protein: 2g
- Fat: 2g

Berry Blissful Parfait with Greek Yogurt and Nuts for a Lutein and Zeaxanthin Boost

Prep Time: 10 minutes
Cook Time: 0 minutes
Portion Size: 2

Ingredients:
- 1 cup Greek yogurt
- 1/2 cup mixed berries
- 1/4 cup granola
- 2 tablespoons chopped nuts (almonds or walnuts)
- 1 tablespoon honey

Instructions:
1. Layer Greek yogurt in serving glasses.
2. Top with mixed berries and granola.
3. Sprinkle chopped nuts over the parfait.
4. Drizzle with honey.
5. Serve immediately.

Nutritional Information (per serving):
- Calories: 250
- Carbohydrates: 30g
- Protein: 12g
- Fat: 10g

Savory Spinach and Feta Frittata for a Vision-Enhancing Breakfast

Prep Time: 10 minutes
Cook Time: 20 minutes
Portion Size: 2

Ingredients:

- 4 eggs
- 1 cup spinach, chopped
- 1/4 cup feta cheese, crumbled
- 1/4 cup cherry tomatoes, halved
- 1 tablespoon olive oil
- Salt and pepper to taste

Instructions:

1. Preheat oven to 375°F (190°C).
2. In a bowl, whisk eggs and season with salt and pepper.
3. Heat olive oil in an oven-safe skillet over medium heat.
4. Add spinach and cook until wilted.
5. Pour eggs over spinach and top with feta and cherry tomatoes.
6. Cook on stovetop for 2-3 minutes until edges set.
7. Transfer to oven and bake for 10-12 minutes until fully cooked.

Nutritional Information (per serving):
- Calories: 250
- Carbohydrates: 5g
- Protein: 18g
- Fat: 18g

Hearty Lentil Soup with Carrots and Kale for a Zinc-Rich Eye Health Meal

Prep Time: 15 minutes
Cook Time: 40 minutes
Portion Size: 2

Ingredients:
- 1 cup lentils, rinsed
- 1 carrot, diced
- 1 cup kale, chopped
- 1 onion, chopped
- 2 cloves garlic, minced
- 1 tablespoon olive oil
- 4 cups vegetable broth
- 1 teaspoon cumin
- Salt and pepper to taste

Instructions:

1. Heat olive oil in a large pot over medium heat.
2. Sauté onion, carrot, and garlic until softened.

3. Add lentils, vegetable broth, and cumin.
4. Bring to a boil, then reduce heat and simmer for 25-30 minutes.
5. Stir in kale and cook for an additional 5 minutes.
6. Season with salt and pepper.

Nutritional Information (per serving):
- Calories: 300
- Carbohydrates: 45g
- Protein: 15g
- Fat: 7g

One-Pan Roasted Chicken with Brussels Sprouts and Sweet Potatoes for a Vitamin-Packed Dinner

Prep Time: 15 minutes
Cook Time: 45 minutes
Portion Size: 2

Ingredients:
- 2 chicken breasts
- 1 cup Brussels sprouts, halved
- 1 large sweet potato, diced
- 1 tablespoon olive oil
- 1 teaspoon dried rosemary
- 1 lemon, sliced
- Salt and pepper to taste

Instructions:
1. Preheat oven to 400°F (200°C).
2. Toss Brussels sprouts and sweet potato with olive oil, rosemary, salt, and pepper.
3. Spread vegetables on a baking sheet and place chicken breasts on top.
4. Arrange lemon slices over chicken and vegetables.
5. Bake for 35-45 minutes until chicken is cooked through and vegetables are tender.

Nutritional Information (per serving):
- Calories: 400
- Carbohydrates: 35g
- Protein: 30g
- Fat: 15g

Vibrant Bell Pepper and Black Bean Salad with Avocado for Antioxidant-Rich Eye Fuel

Prep Time: 15 minutes
Cook Time: 0 minutes
Portion Size: 2

Ingredients:
- 1 red bell pepper, diced
- 1 yellow bell pepper, diced
- 1 can black beans, rinsed and drained
- 1 avocado, diced
- 1/4 cup red onion, finely chopped
- 1/4 cup cilantro, chopped
- 2 tablespoons lime juice
- 1 tablespoon olive oil
- Salt and pepper to taste

Instructions:
1. In a large bowl, combine bell peppers, black beans, avocado, red onion, and cilantro.
2. In a small bowl, whisk together lime juice, olive oil, salt, and pepper.
3. Pour dressing over salad and toss gently.
4. Serve immediately.

Nutritional Information (per serving):
- Calories: 350
- Carbohydrates: 45g
- Protein: 10g
- Fat: 18g

Flavorful Turkey and Sweet Potato Chili with Spinach for a Vision-Supporting Feast

Prep Time: 15 minutes
Cook Time: 40 minutes
Portion Size: 2

Ingredients:
- 1/2 pound ground turkey
- 1 sweet potato, diced
- 2 cups spinach, chopped

- 1 can diced tomatoes
- 1 can kidney beans, rinsed and drained
- 1 onion, chopped
- 2 cloves garlic, minced
- 1 tablespoon olive oil
- 1 tablespoon chili powder
- 1 teaspoon cumin
- Salt and pepper to taste

Instructions:
1. Heat olive oil in a large pot over medium heat.
2. Add onion and garlic, sauté until softened.
3. Add ground turkey and cook until browned.
4. Stir in sweet potato, diced tomatoes, kidney beans, chili powder, cumin, salt, and pepper.
5. Bring to a boil, then reduce heat and simmer for 25-30 minutes until sweet potatoes are tender.
6. Stir in spinach and cook for an additional 5 minutes.

Nutritional Information (per serving):
- Calories: 450
- Carbohydrates: 50g
- Protein: 30g
- Fat: 15g

Nourishing Sweet Potato and Black Bean Burgers with Avocado Crema for Eye Health

Prep Time: 20 minutes
Cook Time: 25 minutes
Portion Size: 2

Ingredients:
- 1 sweet potato, cooked and mashed
- 1 can black beans, rinsed and mashed
- 1/4 cup breadcrumbs
- 1/4 cup red onion, finely chopped
- 1 teaspoon cumin
- 1 teaspoon chili powder
- 1 avocado, mashed
- 1/4 cup Greek yogurt
- 1 tablespoon lime juice
- Salt and pepper to taste

Instructions:

1. Preheat oven to 375°F (190°C).
2. In a bowl, combine sweet potato, black beans, breadcrumbs, red onion, cumin, chili powder, salt, and pepper.
3. Form mixture into patties and place on a baking sheet.
4. Bake for 20-25 minutes, flipping halfway through.
5. In a small bowl, mix mashed avocado, Greek yogurt, lime juice, salt, and pepper to make the crema.
6. Serve burgers with avocado crema on top.

Nutritional Information (per serving):
- Calories: 400
- Carbohydrates: 60g
- Protein: 15g
- Fat: 12g

Revitalizing Carrot and Ginger Soup with Coconut Milk for a Vitamin A-Rich Delight

Prep Time: 15 minutes
Cook Time: 30 minutes
Portion Size: 2

Ingredients:
- 4 carrots, peeled and chopped
- 1 onion, chopped
- 2 cloves garlic, minced
- 1 tablespoon ginger, grated
- 1 can coconut milk
- 2 cups vegetable broth
- 1 tablespoon olive oil
- Salt and pepper to taste

Instructions:

1. Heat olive oil in a pot over medium heat.
2. Sauté onion, garlic, and ginger until fragrant.
3. Add carrots and cook for 5 minutes.
4. Pour in vegetable broth and bring to a boil.
5. Reduce heat and simmer for 20 minutes until carrots are tender.
6. Blend soup until smooth.
7. Stir in coconut milk and season with salt and pepper.
8. Serve warm.

Nutritional Information (per serving):

- Calories: 350
- Carbohydrates: 40g
- Protein: 5g
- Fat: 20g

Eye-Opening Berry and Nut Oatmeal with Chia Seeds for Omega-3 and Fiber

Prep Time: 10 minutes
Cook Time: 10 minutes
Portion Size: 2

Ingredients:
- 1 cup rolled oats
- 2 cups almond milk
- 1/2 cup mixed berries
- 2 tablespoons chia seeds
- 1/4 cup chopped nuts (almonds or walnuts)
- 1 tablespoon honey

Instructions:
1. In a pot, bring almond milk to a boil.
2. Add oats and reduce heat, simmering for 5-7 minutes until thickened.
3. Stir in chia seeds and cook for an additional 2 minutes.
4. Serve oatmeal topped with mixed berries, nuts, and a drizzle of honey.

Nutritional Information (per serving):
- Calories: 300
- Carbohydrates: 45g
- Protein: 8g
- Fat: 12g

Colorful Rainbow Salad with Mixed Greens, Bell Peppers, and Sunflower Seeds for Antioxidant Power

Prep Time: 15 minutes
Cook Time: 0 minutes

Portion Size: 2

Ingredients:
- 2 cups mixed greens
- 1/4 cup red bell pepper, sliced
- 1/4 cup yellow bell pepper, sliced
- 1/4 cup cherry tomatoes, halved
- 1/4 cup shredded carrots
- 2 tablespoons sunflower seeds
- 1/4 cup balsamic vinaigrette

Instructions:
1. In a large bowl, combine mixed greens, bell peppers, cherry tomatoes, and shredded carrots.
2. Sprinkle with sunflower seeds.
3. Drizzle with balsamic vinaigrette and toss gently.
4. Serve immediately.

Nutritional Information (per serving):
- Calories: 200
- Carbohydrates: 20g
- Protein: 5g
- Fat: 12g

Satisfying Salmon Cakes with Dill Yogurt Sauce for Omega-3 and Protein

Prep Time: 15 minutes
Cook Time: 20 minutes
Portion Size: 2

Ingredients:
- 1 can salmon, drained and flaked
- 1/4 cup breadcrumbs
- 1 egg, beaten
- 1/4 cup green onion, chopped
- 1 tablespoon dill, chopped
- 1 tablespoon lemon juice
- 1/4 cup Greek yogurt
- 1 tablespoon olive oil
- Salt and pepper to taste

Instructions:
1. In a bowl, mix salmon, breadcrumbs, egg, green onion, half of the dill, lemon juice, salt, and pepper.
2. Form mixture into patties.

3. Heat olive oil in a skillet over medium heat.
4. Cook patties for 4-5 minutes on each side until golden brown.
5. In a small bowl, combine Greek yogurt and remaining dill to make the sauce.
6. Serve salmon cakes with dill yogurt sauce.

Nutritional Information (per serving):
- Calories: 300
- Carbohydrates: 15g
- Protein: 25g
- Fat: 15g

Energizing Trail Mix with Nuts, Seeds, and Dried Fruit for Eye-Healthy Snacking

Prep Time: 5 minutes
Cook Time: 0 minutes
Portion Size: 2

Ingredients:
- 1/4 cup almonds
- 1/4 cup walnuts
- 1/4 cup pumpkin seeds
- 1/4 cup dried cranberries
- 1/4 cup raisins

Instructions:
1. In a bowl, combine almonds, walnuts, pumpkin seeds, dried cranberries, and raisins.
2. Mix well.
3. Store in an airtight container.
4. Serve as a snack.

Nutritional Information (per serving):
- Calories: 250
- Carbohydrates: 30g
- Protein: 6g
- Fat: 15g

Refreshing Watermelon and Feta Salad with Mint for a Hydrating and Nutrient-Rich Treat

Prep Time: 10 minutes
Cook Time: 0 minutes
Portion Size: 2

Ingredients:
- 2 cups watermelon, cubed
- 1/4 cup feta cheese, crumbled
- 2 tablespoons fresh mint, chopped
- 1 tablespoon lime juice

Instructions:
1. In a bowl, combine watermelon, feta cheese, and mint.
2. Drizzle with lime juice and toss gently.
3. Serve chilled.

Nutritional Information (per serving):
- Calories: 150
- Carbohydrates: 18g
- Protein: 4g
- Fat: 7g

Wholesome Chicken Stir-Fry with Broccoli and Carrots for a Vision-Boosting Meal

Prep Time: 15 minutes
Cook Time: 15 minutes
Portion Size: 2

Ingredients:
- 2 chicken breasts, sliced
- 1 cup broccoli florets
- 1 carrot, sliced
- 1 bell pepper, sliced
- 2 tablespoons soy sauce
- 1 tablespoon olive oil
- 2 cloves garlic, minced
- 1 tablespoon ginger, grated

Instructions:
1. Heat olive oil in a large skillet over medium-high heat.
2. Add garlic and ginger, sauté until fragrant.
3. Add chicken and cook until browned.
4. Add broccoli, carrot, and bell pepper, stirring frequently.
5. Pour in soy sauce and cook until vegetables are tender and chicken is cooked through.

6. Serve immediately.

Nutritional Information (per serving):
- Calories: 350
- Carbohydrates: 20g
- Protein: 30g
- Fat: 12g

Creamy Avocado and Spinach Smoothie with Banana for a Lutein-Rich Boost

Prep Time: 10 minutes
Cook Time: 0 minutes
Portion Size: 2

Ingredients:
- 1 avocado, pitted and peeled
- 1 banana
- 1 cup spinach
- 1 cup almond milk
- 1 tablespoon honey
- 1/2 cup ice

Instructions:
1. Combine avocado, banana, spinach, almond milk, honey, and ice in a blender.
2. Blend until smooth.
3. Serve immediately.

Nutritional Information (per serving):
- Calories: 250
- Carbohydrates: 35g
- Protein: 4g
- Fat: 12g

Spicy Black Bean and Corn Salad with Cilantro Lime Dressing for Antioxidant Protection

Prep Time: 15 minutes

Cook Time: 0 minutes
Portion Size: 2

Ingredients:
- 1 can black beans, rinsed and drained
- 1 cup corn kernels
- 1/4 cup red bell pepper, diced
- 1/4 cup red onion, diced
- 1/4 cup cilantro, chopped
- 2 tablespoons lime juice
- 1 tablespoon olive oil
- 1/2 teaspoon chili powder
- Salt and pepper to taste

Instructions:
1. In a large bowl, combine black beans, corn, bell pepper, red onion, and cilantro.
2. In a small bowl, whisk together lime juice, olive oil, chili powder, salt, and pepper.
3. Pour dressing over salad and toss gently.
4. Serve immediately.

Nutritional Information (per serving):
- Calories: 200
- Carbohydrates: 35g
- Protein: 8g
- Fat: 6g

Comforting Chicken Noodle Soup with Carrots and Celery for a Nourishing and Hydrating Meal

Prep Time: 15 minutes
Cook Time: 30 minutes
Portion Size: 2

Ingredients:
- 1 chicken breast, diced
- 1 cup egg noodles
- 2 carrots, sliced
- 2 celery stalks, sliced
- 1 onion, chopped
- 2 cloves garlic, minced
- 4 cups chicken broth
- 1 tablespoon olive oil
- Salt and pepper to taste

Instructions:

1. Heat olive oil in a large pot over medium heat.
2. Sauté onion, garlic, carrots, and celery until softened.
3. Add chicken and cook until browned.
4. Pour in chicken broth and bring to a boil.
5. Add egg noodles and cook for 8-10 minutes until tender.
6. Season with salt and pepper.
7. Serve warm.

Nutritional Information (per serving):

- Calories: 300
- Carbohydrates: 30g
- Protein: 25g
- Fat: 10g

Foods Rich in Antioxidants for Eye Health

1. **Carrots**: Packed with beta-carotene, which the body converts into vitamin A, essential for maintaining good vision and preventing night blindness.
2. **Spinach**: A rich source of lutein and zeaxanthin, antioxidants that protect the eyes from harmful light and reduce the risk of chronic eye diseases.
3. **Blueberries**: Contain high levels of antioxidants, particularly vitamin C and anthocyanins, which help reduce inflammation and protect the eyes from oxidative stress.
4. **Kale**: Similar to spinach, kale is loaded with lutein and zeaxanthin, along with vitamins C and E, which help safeguard eye tissues from damage caused by free radicals.
5. **Sweet Potatoes**: High in beta-carotene and vitamin E, sweet potatoes help improve vision and protect eyes from age-related damage.
6. **Red Bell Peppers**: These are high in vitamin C, which supports the health of blood vessels in the eyes and may reduce the risk of cataracts.
7. **Oranges**: Rich in vitamin C, oranges help maintain the health of the blood vessels in the eyes and may reduce the risk of cataracts and age-related macular degeneration.
8. **Almonds**: Packed with vitamin E, almonds help protect the eyes from free radical damage and slow down age-related macular degeneration.
9. **Avocados**: Contain lutein and zeaxanthin, as well as vitamins C and E, making them excellent for reducing the risk of cataracts and macular degeneration.
10. **Tomatoes**: High in lycopene and vitamin C, tomatoes provide powerful antioxidants that protect the eyes from light-induced damage and oxidative stress.

Eye Exercises and Relaxation Techniques

1. **Palming**: Rub your hands together to generate heat, then gently place them over your closed eyes without pressing. The warmth and darkness help relax the eye muscles and relieve strain. Practice for 2-3 minutes several times a day.
2. **Focus Shifting**: Hold your thumb about 10 inches away from your face and focus on it for 15 seconds. Then shift your focus to an object about 20 feet away for another 15 seconds. Repeat this process 10 times to improve your focusing ability and reduce eye fatigue.
3. **Figure Eight**: Imagine a large figure eight on the floor about 10 feet in front of you. Trace the figure slowly with your eyes, first in one direction for a minute, then in the opposite direction. This exercise helps enhance eye flexibility and control.
4. **20-20-20 Rule**: To prevent digital eye strain, every 20 minutes, take a 20-second break to look at something 20 feet away. This simple practice reduces eye fatigue and promotes better eye health.
5. **Eye Rolling**: Close your eyes and slowly roll them in a circular motion. Do this clockwise for a minute and then counterclockwise for another minute. Eye rolling can help relieve tension and increase blood circulation to the eyes.
6. **Blinking**: Blinking lubricates the eyes and can help reduce dryness and irritation. Make a conscious effort to blink every few seconds, especially when working on a computer or reading.
7. **Near and Far Focusing**: Hold your thumb 10 inches from your face and focus on it. Then, shift your focus to an object 10-20 feet away. Alternate between near and far focus 10 times to strengthen the eye muscles and improve flexibility.

8. **Zooming**: Hold your arm outstretched with your thumb up. Focus on your thumb as you slowly bring it closer to your nose. Once it's about 3 inches away, slowly extend your arm back out. Repeat this 10 times to improve focus and coordination.
9. **Pencil Push-Ups**: Hold a pencil at arm's length and slowly move it towards your nose, keeping your eyes focused on the pencil. Once you see it double, move it back out. Repeat this exercise 10 times to improve convergence and focusing ability.
10. **Eye Massage**: Using your fingertips, gently massage your temples in a circular motion for about a minute. Then move to the area above your eyebrows and finally under your eyes. This helps improve blood circulation and relaxes the muscles around the eyes.

Natural Hair Care Recipes for Healthy, Strong, and Radiant Hair

Aloe Vera and Coconut Oil Deep Conditioning Hair Mask

Prep Time: 5 minutes

Ingredients:
- 2 tablespoons aloe vera gel
- 2 tablespoons coconut oil

Instructions:
1. Mix aloe vera gel and coconut oil until well blended.
2. Apply the mixture to damp hair, focusing on the ends.
3. Leave on for 30 minutes.
4. Rinse thoroughly with warm water.

Avocado and Honey Nourishing Hair Treatment

Prep Time: 5 minutes

Ingredients:
- 1 ripe avocado
- 2 tablespoons honey

Instructions:
1. Mash the avocado until smooth.
2. Mix in the honey until fully combined.
3. Apply to damp hair, focusing on the scalp and ends.
4. Leave on for 20 minutes.
5. Rinse thoroughly with warm water.

Apple Cider Vinegar Rinse for Shiny Hair

Prep Time: 2 minutes

Ingredients:
- 2 tablespoons apple cider vinegar
- 1 cup water

Instructions:
1. Mix apple cider vinegar with water in a bowl.

2. After shampooing, pour the mixture over your hair.
3. Leave it on for 5 minutes.
4. Rinse thoroughly with cool water.

Rosemary and Lavender Hair Growth Serum

Prep Time: 5 minutes

Ingredients:
- 2 tablespoons jojoba oil
- 5 drops rosemary essential oil
- 5 drops lavender essential oil

Instructions:
1. Mix all ingredients in a small bottle.
2. Massage a few drops into the scalp.
3. Leave on overnight.
4. Wash hair in the morning.

Banana and Olive Oil Moisturizing Hair Mask

Prep Time: 5 minutes

Ingredients:
- 1 ripe banana
- 2 tablespoons olive oil

Instructions:
1. Mash the banana until smooth.
2. Mix in the olive oil until well combined.
3. Apply to damp hair, focusing on dry areas.
4. Leave on for 30 minutes.
5. Rinse thoroughly with warm water.

Yogurt and Egg Protein Treatment for Stronger Hair

Prep Time: 5 minutes

Ingredients:
- 1 egg
- 2 tablespoons plain yogurt

Instructions:
1. Beat the egg in a bowl.
2. Mix in the yogurt until well combined.
3. Apply to clean, damp hair.
4. Leave on for 20 minutes.
5. Rinse thoroughly with cool water.

Green Tea and Peppermint Scalp Detox

Prep Time: 10 minutes

Ingredients:
- 1 cup brewed green tea, cooled
- 5 drops peppermint essential oil

Instructions:
1. Brew green tea and allow it to cool.
2. Add peppermint essential oil to the tea.
3. Pour the mixture over your scalp after shampooing.
4. Leave on for 5 minutes.
5. Rinse thoroughly with cool water.

Castor Oil and Jojoba Oil Split Ends Repair Serum

Prep Time: 2 minutes

Ingredients:
- 1 tablespoon castor oil
- 1 tablespoon jojoba oil

Instructions:
1. Mix castor oil and jojoba oil in a small bottle.
2. Apply a few drops to the ends of your hair.
3. Leave on overnight.
4. Wash hair in the morning.

Chamomile and Lemon Lightening Hair Rinse

Prep Time: 10 minutes

Ingredients:
- 1 cup brewed chamomile tea, cooled
- Juice of 1 lemon

Instructions:
1. Brew chamomile tea and allow it to cool.
2. Mix in the lemon juice.
3. Pour the mixture over your hair after shampooing.
4. Leave on for 10 minutes.
5. Rinse thoroughly with cool water.

Flaxseed Gel for Natural Hair Styling

Prep Time: 15 minutes

Ingredients:
- 1/4 cup flaxseeds
- 2 cups water

Instructions:
1. Boil water in a pot.
2. Add flaxseeds and simmer for 10 minutes, stirring occasionally.
3. Strain the mixture through a fine mesh sieve.
4. Allow the gel to cool before use.
5. Apply a small amount to damp hair and style as desired.

Henna and Amla Strengthening Hair Pack

Prep Time: 10 minutes (plus soaking time)

Ingredients:
- 1/2 cup henna powder
- 1/4 cup amla powder
- Warm water as needed

Instructions:

1. Mix henna and amla powders in a bowl.
2. Add warm water to form a thick paste.
3. Cover and let it sit for 2-3 hours.
4. Apply the paste to clean, damp hair.
5. Leave on for 1-2 hours.
6. Rinse thoroughly with warm water.

Hibiscus and Coconut Milk Hair Thickening Mask

Prep Time: 5 minutes

Ingredients:
- 1/4 cup hibiscus powder
- 1/2 cup coconut milk

Instructions:
1. Mix hibiscus powder and coconut milk until smooth.
2. Apply to clean, damp hair.
3. Leave on for 30 minutes.
4. Rinse thoroughly with warm water.

Onion Juice and Garlic Anti-Hair Loss Treatment

Prep Time: 10 minutes

Ingredients:
- 1 onion
- 3 cloves garlic
- 2 tablespoons olive oil

Instructions:
1. Blend the onion and garlic to extract the juice.
2. Mix the juice with olive oil.
3. Apply the mixture to the scalp.
4. Leave on for 30 minutes.
5. Rinse thoroughly with warm water.

Shea Butter and Argan Oil Frizz Control Cream

Prep Time: 5 minutes

Ingredients:
- 2 tablespoons shea butter
- 1 tablespoon argan oil

Instructions:
1. Melt shea butter in a double boiler.
2. Remove from heat and mix in argan oil.
3. Allow to cool and solidify.
4. Apply a small amount to damp or dry hair.
5. Style as desired.

Baking Soda and Water Clarifying Shampoo

Prep Time: 2 minutes

Ingredients:
- 1 tablespoon baking soda
- 1 cup water

Instructions:
1. Mix baking soda and water in a bowl.
2. Apply the mixture to wet hair and massage into the scalp.
3. Rinse thoroughly with warm water.
4. Follow with a conditioner.

Aloe Vera and Neem Scalp Soothing Gel

Prep Time: 5 minutes

Ingredients:
- 2 tablespoons aloe vera gel
- 1 teaspoon neem oil

Instructions:
1. Mix aloe vera gel and neem oil in a bowl.

2. Apply the mixture to the scalp.
3. Leave on for 20 minutes.
4. Rinse thoroughly with warm water.

Oatmeal and Almond Milk Hydrating Hair Mask

Prep Time: 5 minutes

Ingredients:
- 1/4 cup oatmeal
- 1/2 cup almond milk

Instructions:
1. Blend oatmeal into a fine powder.
2. Mix the oatmeal powder with almond milk until smooth.
3. Apply to clean, damp hair.
4. Leave on for 30 minutes.
5. Rinse thoroughly with warm water.

Lemon Juice and Honey Anti-Dandruff Treatment

Prep Time: 5 minutes

Ingredients:
- 2 tablespoons lemon juice
- 2 tablespoons honey

Instructions:
1. Mix lemon juice and honey in a bowl.
2. Apply the mixture to the scalp.
3. Leave on for 20 minutes.
4. Rinse thoroughly with warm water.

Fenugreek and Yogurt Hair Growth Mask

Prep Time: 10 minutes

Ingredients:
- 2 tablespoons fenugreek seeds
- 1/2 cup plain yogurt

Instructions:
1. Soak fenugreek seeds in water overnight.
2. Grind the soaked seeds into a paste.
3. Mix the fenugreek paste with yogurt.
4. Apply to clean, damp hair.
5. Leave on for 30 minutes.
6. Rinse thoroughly with warm water.

Rice Water Rinse for Strengthening Hair

Prep Time: 5 minutes (plus soaking time)

Ingredients:
- 1/2 cup rice
- 2 cups water

Instructions:
1. Rinse the rice thoroughly and then soak it in water for 30 minutes.
2. Strain the rice, collecting the water in a bowl.
3. After shampooing, pour the rice water over your hair.
4. Leave on for 10 minutes.
5. Rinse thoroughly with cool water.

Tips for Proper Scalp Hygiene

Maintaining a healthy scalp is crucial for promoting strong and radiant hair. Here are some tips to help you achieve optimal scalp hygiene:

1. **Regular Cleansing:** Wash your hair regularly with a gentle shampoo to remove dirt, oil, and product buildup. Aim to wash your hair 2-3 times a week, or more frequently if your scalp tends to get oily.
2. **Scalp Massage:** During washing, gently massage your scalp with your fingertips. This stimulates blood circulation, promotes relaxation, and helps in distributing natural oils.
3. **Avoid Harsh Chemicals:** Choose shampoos and conditioners free from sulfates, parabens, and other harsh chemicals that can strip your scalp of its natural oils and cause irritation.
4. **Balanced Diet:** Maintain a healthy diet rich in vitamins and minerals. Nutrients like vitamins A, C, D, E, and B-complex, as well as minerals like zinc and iron, play a vital role in scalp health.
5. **Stay Hydrated:** Drink plenty of water to keep your scalp and hair hydrated from within. Dehydration can lead to a dry and flaky scalp.
6. **Protect from Sun Exposure:** Wear a hat or use hair products with UV protection when spending extended periods in the sun to prevent scalp sunburn and damage.
7. **Avoid Overuse of Styling Products:** Limit the use of hair sprays, gels, and other styling products that can build up on your scalp. If you use these products, ensure to wash your hair thoroughly.
8. **Keep Hair Tools Clean:** Regularly clean your hairbrushes, combs, and other hair tools to prevent the transfer of dirt and oil back to your scalp.
9. **Manage Stress:** High stress levels can affect your scalp health. Practice stress-relieving activities such as yoga, meditation, or regular exercise to maintain overall well-being.
10. **Consult a Dermatologist:** If you experience persistent scalp issues like severe dandruff, itchiness, or hair loss, consult a dermatologist for proper diagnosis and treatment.

Natural Oral Health Recipes for Healthy Teeth and Gums

Baking Soda and Coconut Oil Whitening Toothpaste

Prep Time: 5 minutes
Portion Size: 2

Ingredients:
- 2 tablespoons baking soda
- 2 tablespoons coconut oil

Instructions:
1. Mix baking soda and coconut oil until a smooth paste forms.
2. Use a pea-sized amount to brush your teeth for 2 minutes.
3. Rinse thoroughly with water.

Aloe Vera and Peppermint Mouthwash

Prep Time: 5 minutes
Portion Size: 2

Ingredients:
- 1 cup aloe vera juice
- 5 drops peppermint essential oil

Instructions:
1. Combine aloe vera juice and peppermint oil in a bottle.
2. Shake well before each use.
3. Swish a small amount in your mouth for 30 seconds, then spit out.

Turmeric and Coconut Oil Gum Healing Paste

Prep Time: 5 minutes
Portion Size: 2

Ingredients:
- 1 teaspoon turmeric powder
- 1 tablespoon coconut oil

Instructions:
1. Mix turmeric powder and coconut oil until well blended.

2. Apply a small amount to your gums and let sit for 10 minutes.
3. Rinse thoroughly with water.

Clove Oil and Sea Salt Sore Gum Relief Rinse

Prep Time: 5 minutes
Portion Size: 2

Ingredients:
- 1 cup warm water
- 1 teaspoon sea salt
- 5 drops clove essential oil

Instructions:
1. Dissolve sea salt in warm water.
2. Add clove essential oil and mix well.
3. Swish the solution in your mouth for 30 seconds, then spit out.

Activated Charcoal and Bentonite Clay Detox Tooth Powder

Prep Time: 5 minutes
Portion Size: 2

Ingredients:
- 1 tablespoon activated charcoal powder
- 1 tablespoon bentonite clay

Instructions:
1. Mix activated charcoal powder and bentonite clay in a small bowl.
2. Dip a damp toothbrush into the powder and brush gently for 2 minutes.
3. Rinse thoroughly with water.

Green Tea and Xylitol Fresh Breath Mouthwash

Prep Time: 10 minutes
Portion Size: 2

Ingredients:
- 1 cup brewed green tea, cooled
- 1 tablespoon xylitol

Instructions:
1. Brew green tea and let it cool.
2. Stir in xylitol until dissolved.
3. Swish a small amount in your mouth for 30 seconds, then spit out.

Myrrh and Echinacea Anti-Inflammatory Gum Gel

Prep Time: 5 minutes
Portion Size: 2

Ingredients:
- 1 tablespoon myrrh powder
- 1 tablespoon echinacea tincture

Instructions:
1. Mix myrrh powder and echinacea tincture to form a gel.
2. Apply a small amount to your gums and let sit for 10 minutes.
3. Rinse thoroughly with water.

Cranberry and Vitamin C Plaque-Reducing Mouth Rinse

Prep Time: 5 minutes
Portion Size: 2

Ingredients:
- 1 cup cranberry juice
- 1 teaspoon vitamin C powder

Instructions:
1. Mix cranberry juice and vitamin C powder until dissolved.
2. Swish a small amount in your mouth for 30 seconds, then spit out.

Olive Oil and Lemon Oil Pulling Solution

Prep Time: 5 minutes
Portion Size: 2

Ingredients:
- 1/4 cup olive oil
- 5 drops lemon essential oil

Instructions:
1. Combine olive oil and lemon oil in a small bowl.
2. Swish a tablespoon of the mixture in your mouth for 15-20 minutes.
3. Spit out and rinse thoroughly with water.

Neem and Tea Tree Oil Antibacterial Mouthwash

Prep Time: 5 minutes
Portion Size: 2

Ingredients:
- 1 cup water
- 1 teaspoon neem oil
- 5 drops tea tree oil

Instructions:
1. Mix water, neem oil, and tea tree oil in a bottle.
2. Shake well before each use.
3. Swish a small amount in your mouth for 30 seconds, then spit out.

Sage and Sea Salt Teeth Strengthening Rinse

Prep Time: 5 minutes
Portion Size: 2

Ingredients:
- 1 cup warm water
- 1 teaspoon sea salt
- 1 tablespoon dried sage leaves

Instructions:
1. Steep dried sage leaves in warm water for 5 minutes.
2. Strain and add sea salt to the infused water.
3. Swish the mixture in your mouth for 30 seconds, then spit out.

Hydrogen Peroxide and Baking Soda Whitening Gel

Prep Time: 5 minutes
Portion Size: 2

Ingredients:
- 1 tablespoon hydrogen peroxide
- 1 tablespoon baking soda

Instructions:
1. Mix hydrogen peroxide and baking soda to form a gel.
2. Apply a small amount to your teeth with a toothbrush.
3. Let sit for 2 minutes, then rinse thoroughly with water.

Cinnamon and Honey Antibacterial Mouth Rinse

Prep Time: 5 minutes
Portion Size: 2

Ingredients:
- 1 cup warm water
- 1 teaspoon cinnamon powder
- 1 tablespoon honey

Instructions:
1. Dissolve cinnamon powder and honey in warm water.
2. Swish a small amount in your mouth for 30 seconds, then spit out.

Propolis and Aloe Vera Healing Mouth Gel

Prep Time: 5 minutes
Portion Size: 2

Ingredients:
- 1 tablespoon propolis tincture
- 1 tablespoon aloe vera gel

Instructions:
1. Mix propolis tincture and aloe vera gel in a small bowl.
2. Apply a small amount to the affected areas in your mouth.
3. Let sit for 10 minutes, then rinse thoroughly with water.

Eucalyptus Oil and Peppermint Oil Fresh Breath Spray

Prep Time: 5 minutes
Portion Size: 2

Ingredients:
- 1/4 cup water
- 5 drops eucalyptus essential oil
- 5 drops peppermint essential oil

Instructions:
1. Mix water, eucalyptus oil, and peppermint oil in a small spray bottle.
2. Shake well before each use.
3. Spray a small amount into your mouth for fresh breath.

Licorice Root and Spearmint Cavity Prevention Toothpaste

Prep Time: 5 minutes
Portion Size: 2

Ingredients:
- 1 tablespoon licorice root powder
- 1 tablespoon spearmint powder
- 2 tablespoons coconut oil

Instructions:
1. Mix licorice root powder, spearmint powder, and coconut oil to form a paste.
2. Use a pea-sized amount to brush your teeth for 2 minutes.
3. Rinse thoroughly with water.

Chamomile and Clove Gum Soothing Rinse

Prep Time: 5 minutes
Portion Size: 2

Ingredients:
- 1 cup warm water
- 1 tablespoon dried chamomile flowers
- 5 drops clove essential oil

Instructions:
1. Steep dried chamomile flowers in warm water for 5 minutes.
2. Strain and add clove essential oil to the infused water.
3. Swish the mixture in your mouth for 30 seconds, then spit out.

Calcium Powder and Xylitol Remineralizing Tooth Powder

Prep Time: 5 minutes
Portion Size: 2

Ingredients:
- 2 tablespoons calcium powder
- 1 tablespoon xylitol
- 1 tablespoon baking soda

Instructions:
1. Mix calcium powder, xylitol, and baking soda in a small bowl.
2. Dip a damp toothbrush into the powder and brush gently for 2 minutes.
3. Rinse thoroughly with water.

Basil and Fennel Natural Breath Freshener

Prep Time: 5 minutes
Portion Size: 2

Ingredients:
- 1 tablespoon dried basil leaves
- 1 tablespoon fennel seeds
- 1 cup boiling water

Instructions:
1. Steep dried basil leaves and fennel seeds in boiling water for 5 minutes.
2. Strain the mixture and let it cool.

3. Swish a small amount in your mouth for 30 seconds, then spit out.

Apple Cider Vinegar and Baking Soda Natural Tooth Cleaner

Prep Time: 5 minutes
Portion Size: 2

Ingredients:
- 2 tablespoons apple cider vinegar
- 1 tablespoon baking soda

Instructions:
1. Mix apple cider vinegar and baking soda to form a paste.
2. Use a pea-sized amount to brush your teeth for 2 minutes.
3. Rinse thoroughly with water.

Tips for Proper Oral Hygiene

1. **Brush Regularly:** Brush your teeth at least twice a day using a fluoride toothpaste. Ensure to brush for at least two minutes, covering all surfaces of your teeth.
2. **Floss Daily:** Flossing removes plaque and food particles from between your teeth and under the gumline, where your toothbrush can't reach. Make it a daily habit to prevent gum disease and cavities.
3. **Use Mouthwash:** Rinse your mouth with an antibacterial mouthwash to reduce plaque, prevent gingivitis, and maintain fresh breath. Choose a mouthwash that suits your specific oral health needs.
4. **Maintain a Healthy Diet:** Limit sugary and acidic foods and drinks, as they can erode tooth enamel and lead to cavities. Incorporate plenty of fruits, vegetables, whole grains, and dairy products into your diet for overall oral health.
5. **Stay Hydrated:** Drink plenty of water throughout the day to keep your mouth hydrated and wash away food particles and bacteria.
6. **Avoid Tobacco:** Smoking and chewing tobacco can cause serious oral health problems, including gum disease, tooth decay, and oral cancer. Avoiding tobacco products is crucial for maintaining a healthy mouth.
7. **Regular Dental Check-ups:** Visit your dentist regularly for check-ups and professional cleanings. Regular dental visits help detect and treat oral health issues early.
8. **Replace Your Toothbrush:** Change your toothbrush or toothbrush head every three to four months, or sooner if the bristles are frayed. A worn-out toothbrush is less effective at cleaning your teeth.
9. **Clean Your Tongue:** Use a tongue scraper or your toothbrush to clean your tongue daily. This helps remove bacteria and freshen your breath.
10. **Wear a Mouthguard:** If you play sports or grind your teeth at night, wear a mouthguard to protect your teeth from damage.

Natural Recipes for Women's Health

Hormone-Balancing Flaxseed and Berry Smoothie

Prep Time: 5 minutes
Portion Size: 2

Ingredients:
- 2 tablespoons flaxseeds
- 1 cup mixed berries (fresh or frozen)
- 1 banana
- 1 cup almond milk
- 1 tablespoon honey (optional)

Instructions:
1. Combine all ingredients in a blender.
2. Blend until smooth.
3. Pour into glasses and serve immediately.

Nutritional Information (per serving):
- Calories: 250
- Carbohydrates: 45g
- Protein: 5g
- Fat: 8g

Iron-Rich Beet and Spinach Salad

Prep Time: 10 minutes
Portion Size: 2

Ingredients:
- 2 medium beets, cooked and sliced
- 4 cups fresh spinach
- 1/4 cup crumbled feta cheese
- 2 tablespoons pumpkin seeds
- 2 tablespoons olive oil
- 1 tablespoon balsamic vinegar
- Salt and pepper to taste

Instructions:
1. Arrange spinach on plates.
2. Top with beet slices, feta cheese, and pumpkin seeds.
3. Drizzle with olive oil and balsamic vinegar.
4. Season with salt and pepper.

Nutritional Information (per serving):
- Calories: 200
- Carbohydrates: 15g
- Protein: 8g
- Fat: 14g

Anti-Inflammatory Turmeric and Ginger Tea

Prep Time: 5 minutes
Portion Size: 2

Ingredients:
- 2 cups water
- 1 teaspoon turmeric powder
- 1 teaspoon fresh grated ginger
- 1 tablespoon honey (optional)
- Lemon slices (optional)

Instructions:
1. Bring water to a boil.
2. Add turmeric and ginger, reduce heat, and simmer for 10 minutes.
3. Strain into cups and add honey and lemon if desired.

Nutritional Information (per serving):
- Calories: 30
- Carbohydrates: 8g
- Protein: 0g
- Fat: 0g

Calcium-Boosting Almond and Kale Smoothie

Prep Time: 5 minutes
Portion Size: 2

Ingredients:
- 1 cup almond milk
- 1 cup fresh kale
- 1 banana
- 2 tablespoons almond butter
- 1 tablespoon chia seeds

Instructions:
1. Combine all ingredients in a blender.
2. Blend until smooth.
3. Pour into glasses and serve immediately.

Nutritional Information (per serving):
- Calories: 300
- Carbohydrates: 35g
- Protein: 7g
- Fat: 18g

Menopause Relief Sage and Lemon Balm Tea

Prep Time: 5 minutes
Portion Size: 2

Ingredients:
- 2 cups water
- 1 tablespoon dried sage
- 1 tablespoon dried lemon balm
- Honey (optional)

Instructions:
1. Bring water to a boil.
2. Add sage and lemon balm, reduce heat, and simmer for 5 minutes.
3. Strain into cups and add honey if desired.

Nutritional Information (per serving):
- Calories: 5
- Carbohydrates: 1g
- Protein: 0g
- Fat: 0g

PMS Soothing Chamomile and Raspberry Leaf Tea

Prep Time: 5 minutes
Portion Size: 2

Ingredients:
- 2 cups water
- 1 tablespoon dried chamomile flowers

- 1 tablespoon dried raspberry leaf
- Honey (optional)

Instructions:
1. Bring water to a boil.
2. Add chamomile and raspberry leaf, reduce heat, and simmer for 5 minutes.
3. Strain into cups and add honey if desired.

Nutritional Information (per serving):
- Calories: 5
- Carbohydrates: 1g
- Protein: 0g
- Fat: 0g

Bone Strengthening Sesame and Chia Seed Pudding

Prep Time: 10 minutes
Portion Size: 2

Ingredients:
- 1/4 cup chia seeds
- 1 cup almond milk
- 2 tablespoons sesame seeds
- 1 tablespoon honey
- 1/2 teaspoon vanilla extract

Instructions:
1. Mix all ingredients in a bowl.
2. Let sit for at least 4 hours or overnight.
3. Stir before serving.

Nutritional Information (per serving):
- Calories: 250
- Carbohydrates: 20g
- Protein: 8g
- Fat: 15g

Skin-Glowing Avocado and Cucumber Salad

Prep Time: 10 minutes
Portion Size: 2

Ingredients:
- 1 avocado, diced
- 1 cucumber, sliced
- 1/4 cup red onion, thinly sliced
- 2 tablespoons fresh cilantro, chopped
- 2 tablespoons olive oil
- Juice of 1 lime
- Salt and pepper to taste

Instructions:
1. Combine avocado, cucumber, red onion, and cilantro in a bowl.
2. Drizzle with olive oil and lime juice.
3. Season with salt and pepper and toss gently.

Nutritional Information (per serving):
- Calories: 220
- Carbohydrates: 15g
- Protein: 2g
- Fat: 19g

Stress-Relief Lavender and Valerian Root Tea

Prep Time: 5 minutes
Portion Size: 2

Ingredients:
- 2 cups water
- 1 teaspoon dried lavender flowers
- 1 teaspoon dried valerian root
- Honey (optional)

Instructions:
1. Bring water to a boil.
2. Add lavender and valerian root, reduce heat, and simmer for 5 minutes.
3. Strain into cups and add honey if desired.

Nutritional Information (per serving):
- Calories: 5
- Carbohydrates: 1g
- Protein: 0g
- Fat: 0g

Energy-Boosting Maca and Banana Smoothie

Prep Time: 5 minutes
Portion Size: 2

Ingredients:
- 1 cup almond milk
- 1 banana
- 1 tablespoon maca powder
- 1 tablespoon almond butter
- 1 teaspoon honey (optional)

Instructions:
1. Combine all ingredients in a blender.
2. Blend until smooth.
3. Pour into glasses and serve immediately.

Nutritional Information (per serving):
- Calories: 200
- Carbohydrates: 35g
- Protein: 4g
- Fat: 7g

Immune-Boosting Elderberry and Hibiscus Tea

Prep Time: 10 minutes
Portion Size: 2

Ingredients:
- 2 cups water
- 1 tablespoon dried elderberries
- 1 tablespoon dried hibiscus flowers
- Honey (optional)

Instructions:
1. Bring water to a boil.
2. Add elderberries and hibiscus, reduce heat, and simmer for 10 minutes.
3. Strain into cups and add honey if desired.

Nutritional Information (per serving):
- Calories: 10
- Carbohydrates: 2g
- Protein: 0g

- Fat: 0g

Detoxifying Dandelion and Mint Tea

Prep Time: 5 minutes
Portion Size: 2

Ingredients:
- 2 cups water
- 1 tablespoon dried dandelion root
- 1 tablespoon dried mint leaves
- Honey (optional)

Instructions:
1. Bring water to a boil.
2. Add dandelion root and mint leaves, reduce heat, and simmer for 5 minutes.
3. Strain into cups and add honey if desired.

Nutritional Information (per serving):
- Calories: 5
- Carbohydrates: 1g
- Protein: 0g
- Fat: 0g

Mood-Enhancing Dark Chocolate and Walnut Bites

Prep Time: 10 minutes
Portion Size: 2

Ingredients:
- 1/2 cup dark chocolate chips
- 1/4 cup chopped walnuts
- 1 tablespoon coconut oil

Instructions:
1. Melt dark chocolate chips and coconut oil in a double boiler.
2. Stir in chopped walnuts.
3. Drop spoonfuls of the mixture onto parchment paper.
4. Refrigerate until firm.

Nutritional Information (per serving):

- Calories: 200
- Carbohydrates: 15g
- Protein: 3g
- Fat: 15g

Digestive Health Fennel and Peppermint Tea

Prep Time: 5 minutes
Portion Size: 2

Ingredients:
- 2 cups water
- 1 teaspoon fennel seeds
- 1 teaspoon dried peppermint leaves
- Honey (optional)

Instructions:
1. Bring water to a boil.
2. Add fennel seeds and peppermint leaves, reduce heat, and simmer for 5 minutes.
3. Strain into cups and add honey if desired.

Nutritional Information (per serving):
- Calories: 5
- Carbohydrates: 1g
- Protein: 0g
- Fat: 0g

Uterine Health Red Clover and Nettle Infusion

Prep Time: 5 minutes
Portion Size: 2

Ingredients:
- 2 cups water
- 1 tablespoon dried red clover
- 1 tablespoon dried nettle
- Honey (optional)

Instructions:
1. Bring water to a boil.
2. Add red clover and nettle, reduce heat, and simmer for 5 minutes.

3. Strain into cups and add honey if desired.

Nutritional Information (per serving):
- Calories: 5
- Carbohydrates: 1g
- Protein: 0g
- Fat: 0g

Hair and Nail Strengthening Biotin-Rich Smoothie

Prep Time: 5 minutes
Portion Size: 2

Ingredients:
- 1 cup almond milk
- 1 banana
- 1/2 cup strawberries
- 1 tablespoon almond butter
- 1 tablespoon chia seeds

Instructions:
1. Combine all ingredients in a blender.
2. Blend until smooth.
3. Pour into glasses and serve immediately.

Nutritional Information (per serving):
- Calories: 250
- Carbohydrates: 35g
- Protein: 6g
- Fat: 12g

Heart Health Omega-3 Flaxseed and Walnut Mix

Prep Time: 5 minutes
Portion Size: 2

Ingredients:
- 1/4 cup flaxseeds
- 1/4 cup chopped walnuts
- 2 tablespoons dried cranberries

Instructions:
1. Combine flaxseeds, walnuts, and dried cranberries in a bowl.
2. Mix well and serve as a snack.

Nutritional Information (per serving):
- Calories: 200
- Carbohydrates: 10g
- Protein: 5g
- Fat: 16g

Hydrating Coconut Water and Aloe Vera Drink

Prep Time: 5 minutes
Portion Size: 2

Ingredients:
- 2 cups coconut water
- 1/4 cup aloe vera juice
- 1 tablespoon lemon juice

Instructions:
1. Combine coconut water, aloe vera juice, and lemon juice in a pitcher.
2. Stir well and pour into glasses.
3. Serve chilled.

Nutritional Information (per serving):
- Calories: 30
- Carbohydrates: 7g
- Protein: 0g
- Fat: 0g

Anti-Aging Green Tea and Blueberry Smoothie

Prep Time: 5 minutes
Portion Size: 2

Ingredients:
- 1 cup brewed green tea, cooled
- 1/2 cup blueberries
- 1 banana
- 1 tablespoon honey (optional)

Instructions:

1. Combine green tea, blueberries, banana, and honey in a blender.
2. Blend until smooth.
3. Pour into glasses and serve immediately.

Nutritional Information (per serving):

- Calories: 100
- Carbohydrates: 25g
- Protein: 1g
- Fat: 0g

Reproductive Health Maca and Cinnamon Latte

Prep Time: 5 minutes
Portion Size: 2

Ingredients:

- 2 cups almond milk
- 1 tablespoon maca powder
- 1/2 teaspoon cinnamon
- 1 tablespoon honey (optional)

Instructions:

1. Heat almond milk in a saucepan over medium heat.
2. Whisk in maca powder, cinnamon, and honey until well combined.
3. Pour into mugs and serve warm.

Nutritional Information (per serving):

- Calories: 120
- Carbohydrates: 22g
- Protein: 2g
- Fat: 3g

Herbs for Menstrual Regulation, Menopause, and Fertility

For Menstrual Regulation:

Chasteberry (Vitex): Helps balance hormones and regulate menstrual cycles.

Ginger: Reduces menstrual pain and inflammation.

Dong Quai: Known as the "female ginseng," it helps balance estrogen levels.

Black Cohosh: Relieves symptoms of PMS and promotes menstrual regularity.

For Menopause Relief:

Red Clover: Contains phytoestrogens that help alleviate menopausal symptoms.

Sage: Reduces hot flashes and night sweats.

Wild Yam: Supports hormonal balance and alleviates menopausal discomfort.

Licorice Root: Helps manage hot flashes and hormone-related mood swings.

For Fertility Enhancement:

Maca Root: Increases libido and balances hormones.

Nettle: Rich in vitamins and minerals, it supports reproductive health.

Tribulus: Enhances fertility and promotes ovulation.

Raspberry Leaf: Strengthens the uterine lining and supports overall reproductive health.

30-hour video course, divided into 7 modules, directly from Dr. Barbara O'Neill

Module 1: Discover if genetics cause disease, the role of germs, how to conquer health, and eliminate toxins from the body.

Module 2: Boost your immune system, heal the respiratory organs, decode diabetes, and heal the heart and high blood pressure.

Module 3: Lose weight easily, improve sleep, and learn natural remedies for every need.

Module 4: Dive into water therapy, discover child nutrition, balance your hormones, and maintain strong bones.

Module 5: Understand DNA and the true causes of disease, optimize the body's elimination systems, and manage your sexual and hormonal health.

Module 6: Enjoy pure air, sunlight, and proper rest, and discover the importance of healthy fats and acid-alkaline balance.

Module 7: Exercise effectively, understand the importance of salt, the laws of the mind, and how to maintain gut health.

Natural Recipes for Men's Health

Testosterone-Boosting Walnut and Banana Smoothie

Prep Time: 5 minutes
Portion Size: 2

Ingredients:
- 1 banana
- 1/2 cup walnuts
- 1 cup almond milk
- 1 tablespoon honey (optional)
- 1/2 teaspoon vanilla extract

Instructions:
1. Combine all ingredients in a blender.
2. Blend until smooth.
3. Pour into glasses and serve immediately.

Nutritional Information (per serving):
- Calories: 250
- Carbohydrates: 30g
- Protein: 5g
- Fat: 15g

Heart-Healthy Avocado and Spinach Salad

Prep Time: 10 minutes
Portion Size: 2

Ingredients:
- 1 avocado, diced
- 4 cups fresh spinach
- 1/4 cup cherry tomatoes, halved
- 1/4 cup red onion, thinly sliced
- 2 tablespoons olive oil
- 1 tablespoon balsamic vinegar
- Salt and pepper to taste

Instructions:
1. Arrange spinach on plates.
2. Top with avocado, cherry tomatoes, and red onion.
3. Drizzle with olive oil and balsamic vinegar.
4. Season with salt and pepper and toss gently.

Nutritional Information (per serving):
- Calories: 200
- Carbohydrates: 15g
- Protein: 3g
- Fat: 17g

Anti-Inflammatory Turmeric and Black Pepper Tea

Prep Time: 5 minutes
Portion Size: 2

Ingredients:
- 2 cups water
- 1 teaspoon turmeric powder
- 1/4 teaspoon black pepper
- 1 tablespoon honey (optional)
- Lemon slices (optional)

Instructions:
1. Bring water to a boil.
2. Add turmeric and black pepper, reduce heat, and simmer for 10 minutes.
3. Strain into cups and add honey and lemon if desired.

Nutritional Information (per serving):
- Calories: 15
- Carbohydrates: 4g
- Protein: 0g
- Fat: 0g

Energy-Enhancing Maca and Almond Smoothie

Prep Time: 5 minutes
Portion Size: 2

Ingredients:
- 1 cup almond milk
- 1 banana
- 1 tablespoon maca powder
- 1 tablespoon almond butter
- 1 teaspoon honey (optional)

Instructions:
1. Combine all ingredients in a blender.
2. Blend until smooth.
3. Pour into glasses and serve immediately.

Nutritional Information (per serving):
- Calories: 200
- Carbohydrates: 30g
- Protein: 5g
- Fat: 9g

Prostate Health Pomegranate and Flaxseed Juice

Prep Time: 5 minutes
Portion Size: 2

Ingredients:
- 1 cup pomegranate juice
- 1 tablespoon flaxseed
- 1 tablespoon lemon juice

Instructions:
1. Mix pomegranate juice, flaxseed, and lemon juice in a blender.
2. Blend until smooth.
3. Pour into glasses and serve immediately.

Nutritional Information (per serving):
- Calories: 150
- Carbohydrates: 25g
- Protein: 1g
- Fat: 5g

Muscle Recovery Tart Cherry and Ginger Smoothie

Prep Time: 5 minutes
Portion Size: 2

Ingredients:
- 1 cup tart cherry juice
- 1 banana
- 1 teaspoon grated fresh ginger

- 1/2 cup Greek yogurt

Instructions:
1. Combine all ingredients in a blender.
2. Blend until smooth.
3. Pour into glasses and serve immediately.

Nutritional Information (per serving):
- Calories: 180
- Carbohydrates: 35g
- Protein: 5g
- Fat: 2g

Immune-Boosting Elderberry and Echinacea Tea

Prep Time: 10 minutes
Portion Size: 2

Ingredients:
- 2 cups water
- 1 tablespoon dried elderberries
- 1 tablespoon dried echinacea
- Honey (optional)

Instructions:
1. Bring water to a boil.
2. Add elderberries and echinacea, reduce heat, and simmer for 10 minutes.
3. Strain into cups and add honey if desired.

Nutritional Information (per serving):
- Calories: 10
- Carbohydrates: 3g
- Protein: 0g
- Fat: 0g

Stress-Reducing Ashwagandha and Cinnamon Latte

Prep Time: 5 minutes
Portion Size: 2

Ingredients:

- 2 cups almond milk
- 1 teaspoon ashwagandha powder
- 1/2 teaspoon cinnamon
- 1 tablespoon honey (optional)

Instructions:

1. Heat almond milk in a saucepan over medium heat.
2. Whisk in ashwagandha powder, cinnamon, and honey until well combined.
3. Pour into mugs and serve warm.

Nutritional Information (per serving):

- Calories: 70
- Carbohydrates: 12g
- Protein: 1g
- Fat: 2g

Brain-Boosting Blueberry and Chia Seed Smoothie

Prep Time: 5 minutes
Portion Size: 2

Ingredients:

- 1 cup almond milk
- 1/2 cup blueberries
- 1 banana
- 1 tablespoon chia seeds

Instructions:

1. Combine all ingredients in a blender.
2. Blend until smooth.
3. Pour into glasses and serve immediately.

Nutritional Information (per serving):

- Calories: 150
- Carbohydrates: 30g
- Protein: 3g
- Fat: 4g

Detoxifying Green Tea and Lemon Drink

Prep Time: 5 minutes

Portion Size: 2

Ingredients:
- 2 cups brewed green tea, cooled
- Juice of 1 lemon
- 1 tablespoon honey (optional)

Instructions:
1. Brew green tea and let it cool.
2. Mix in lemon juice and honey.
3. Pour into glasses and serve chilled.

Nutritional Information (per serving):
- Calories: 20
- Carbohydrates: 6g
- Protein: 0g
- Fat: 0g

Bone Strengthening Sesame and Almond Milk

Prep Time: 5 minutes
Portion Size: 2

Ingredients:
- 2 cups almond milk
- 1 tablespoon sesame seeds
- 1 teaspoon honey (optional)

Instructions:
1. Blend almond milk and sesame seeds until smooth.
2. Add honey if desired.
3. Pour into glasses and serve.

Nutritional Information (per serving):
- Calories: 80
- Carbohydrates: 3g
- Protein: 3g
- Fat: 6g

Libido-Enhancing Pumpkin Seed and Cacao Smoothie

Prep Time: 5 minutes
Portion Size: 2

Ingredients:
- 1 cup almond milk
- 1/4 cup pumpkin seeds
- 1 tablespoon raw cacao powder
- 1 banana
- 1 tablespoon honey (optional)

Instructions:
1. Combine all ingredients in a blender.
2. Blend until smooth.
3. Pour into glasses and serve immediately.

Nutritional Information (per serving):
- Calories: 220
- Carbohydrates: 35g
- Protein: 7g
- Fat: 10g

Anti-Anxiety Lavender and Chamomile Tea

Prep Time: 5 minutes
Portion Size: 2

Ingredients:
- 2 cups water
- 1 teaspoon dried lavender flowers
- 1 teaspoon dried chamomile flowers
- Honey (optional)

Instructions:
1. Bring water to a boil.
2. Add lavender and chamomile, reduce heat, and simmer for 5 minutes.
3. Strain into cups and add honey if desired.

Nutritional Information (per serving):
- Calories: 5
- Carbohydrates: 1g
- Protein: 0g
- Fat: 0g

Liver Health Beet and Carrot Juice

Prep Time: 10 minutes
Portion Size: 2

Ingredients:
- 2 medium beets, peeled and chopped
- 2 carrots, peeled and chopped
- 1 apple, cored and chopped
- 1 tablespoon lemon juice

Instructions:
1. Blend beets, carrots, and apple until smooth.
2. Add lemon juice and mix well.
3. Pour into glasses and serve immediately.

Nutritional Information (per serving):
- Calories: 150
- Carbohydrates: 35g
- Protein: 2g
- Fat: 0g

Digestive Aid Fennel and Peppermint Tea

Prep Time: 5 minutes
Portion Size: 2

Ingredients:
- 2 cups water
- 1 teaspoon fennel seeds
- 1 teaspoon dried peppermint leaves
- Honey (optional)

Instructions:
1. Bring water to a boil.
2. Add fennel seeds and peppermint leaves, reduce heat, and simmer for 5 minutes.
3. Strain into cups and add honey if desired.

Nutritional Information (per serving):
- Calories: 5
- Carbohydrates: 1g
- Protein: 0g
- Fat: 0g

Blood Pressure Regulating Hibiscus and Hawthorn Tea

Prep Time: 5 minutes
Portion Size: 2

Ingredients:
- 2 cups water
- 1 tablespoon dried hibiscus flowers
- 1 tablespoon dried hawthorn berries
- Honey (optional)

Instructions:
1. Bring water to a boil.
2. Add hibiscus flowers and hawthorn berries, reduce heat, and simmer for 5 minutes.
3. Strain into cups and add honey if desired.

Nutritional Information (per serving):
- Calories: 10
- Carbohydrates: 2g
- Protein: 0g
- Fat: 0g

Anti-Aging Dark Chocolate and Walnut Bites

Prep Time: 10 minutes
Portion Size: 2

Ingredients:
- 1/2 cup dark chocolate chips
- 1/4 cup chopped walnuts
- 1 tablespoon coconut oil

Instructions:
1. Melt dark chocolate chips and coconut oil in a double boiler.
2. Stir in chopped walnuts.
3. Drop spoonfuls of the mixture onto parchment paper.
4. Refrigerate until firm.

Nutritional Information (per serving):
- Calories: 200

- Carbohydrates: 15g
- Protein: 3g
- Fat: 15g

Hair Growth Rosemary and Nettle Infusion

Prep Time: 10 minutes
Portion Size: 2

Ingredients:
- 2 cups water
- 1 tablespoon dried rosemary
- 1 tablespoon dried nettle

Instructions:
1. Bring water to a boil.
2. Add rosemary and nettle, reduce heat, and simmer for 10 minutes.
3. Strain into cups and serve warm.

Nutritional Information (per serving):
- Calories: 5
- Carbohydrates: 1g
- Protein: 0g
- Fat: 0g

Skin-Clearing Aloe Vera and Lemon Drink

Prep Time: 5 minutes
Portion Size: 2

Ingredients:
- 1/4 cup aloe vera juice
- Juice of 1 lemon
- 2 cups water
- 1 tablespoon honey (optional)

Instructions:
1. Combine aloe vera juice, lemon juice, and water in a pitcher.
2. Stir well and add honey if desired.
3. Pour into glasses and serve chilled.

Nutritional Information (per serving):
- Calories: 20
- Carbohydrates: 6g
- Protein: 0g
- Fat: 0g

Joint Health Turmeric and Pineapple Smoothie

Prep Time: 5 minutes
Portion Size: 2

Ingredients:
- 1 cup pineapple chunks (fresh or frozen)
- 1 cup almond milk
- 1 teaspoon turmeric powder
- 1/2 teaspoon ginger powder
- 1 tablespoon honey (optional)

Instructions:
1. Combine all ingredients in a blender.
2. Blend until smooth.
3. Pour into glasses and serve immediately.

Nutritional Information (per serving):
- Calories: 150
- Carbohydrates: 30g
- Protein: 2g
- Fat: 3g

Herbs for Prostate Health, Fertility, and Libido

For Prostate Health:

Saw Palmetto: Helps reduce symptoms of an enlarged prostate and improves urinary function.

Pygeum: Supports prostate health and reduces inflammation.

Stinging Nettle: Alleviates urinary symptoms associated with prostate enlargement.

Pumpkin Seed: Rich in zinc and supports overall prostate health.

For Fertility Enhancement:

Maca Root: Improves sperm quality and boosts libido.

Tribulus Terrestris: Enhances sperm production and increases testosterone levels.

Ashwagandha: Reduces stress and improves sperm count and motility.

Horny Goat Weed: Boosts testosterone levels and improves erectile function.

For Libido Boost:

Ginseng: Enhances sexual performance and stamina.

Yohimbe: Increases blood flow to the genital area, improving sexual arousal.

Muira Puama: Known as "potency wood," it enhances libido and sexual function.

Damiana: Traditionally used as an aphrodisiac, it helps improve sexual desire and performance.

Natural Recipes for Kidney Health

Kidney Cleansing Watermelon and Mint Juice

Prep Time: 10 minutes
Portion Size: 2

Ingredients:
- 2 cups diced watermelon
- 1 tablespoon fresh mint leaves
- Juice of 1 lime

Instructions:
1. Blend the watermelon, mint leaves, and lime juice until smooth.
2. Strain the mixture to remove pulp, if desired.
3. Pour into glasses and serve chilled.

Nutritional Information (per serving):
- Calories: 70
- Carbohydrates: 18g
- Protein: 1g
- Fat: 0g

Anti-Inflammatory Turmeric and Ginger Tea

Prep Time: 5 minutes
Portion Size: 2

Ingredients:
- 2 cups water
- 1 teaspoon turmeric powder
- 1 teaspoon fresh grated ginger
- 1 tablespoon honey (optional)
- Lemon slices (optional)

Instructions:
1. Bring water to a boil.
2. Add turmeric and ginger, reduce heat, and simmer for 10 minutes.
3. Strain into cups and add honey and lemon if desired.

Nutritional Information (per serving):
- Calories: 15
- Carbohydrates: 4g
- Protein: 0g
- Fat: 0g

Detoxifying Cranberry and Lemon Drink

Prep Time: 5 minutes
Portion Size: 2

Ingredients:
- 1 cup unsweetened cranberry juice
- 1 cup water
- Juice of 1 lemon
- 1 tablespoon honey (optional)

Instructions:
1. Mix cranberry juice, water, lemon juice, and honey in a pitcher.
2. Stir well and pour into glasses.
3. Serve chilled.

Nutritional Information (per serving):
- Calories: 40
- Carbohydrates: 10g
- Protein: 0g
- Fat: 0g

Parsley and Celery Kidney Flush Smoothie

Prep Time: 5 minutes
Portion Size: 2

Ingredients:
- 1 cup fresh parsley
- 2 celery stalks
- 1 apple, cored and chopped
- 1 cup water

Instructions:
1. Combine all ingredients in a blender.
2. Blend until smooth.
3. Pour into glasses and serve immediately.

Nutritional Information (per serving):
- Calories: 50

- Carbohydrates: 12g
- Protein: 1g
- Fat: 0g

Dandelion and Nettle Detox Tea

Prep Time: 10 minutes
Portion Size: 2

Ingredients:
- 2 cups water
- 1 tablespoon dried dandelion root
- 1 tablespoon dried nettle leaves

Instructions:
1. Bring water to a boil.
2. Add dandelion root and nettle leaves, reduce heat, and simmer for 10 minutes.
3. Strain into cups and serve warm.

Nutritional Information (per serving):
- Calories: 5
- Carbohydrates: 1g
- Protein: 0g
- Fat: 0g

Hydrating Cucumber and Lemon Water

Prep Time: 5 minutes
Portion Size: 2

Ingredients:
- 1 cucumber, sliced
- 1 lemon, sliced
- 4 cups water

Instructions:
1. Combine cucumber and lemon slices in a pitcher.
2. Add water and stir well.
3. Refrigerate for at least 1 hour before serving.

Nutritional Information (per serving):

- Calories: 5
- Carbohydrates: 1g
- Protein: 0g
- Fat: 0g

Apple and Carrot Kidney Cleansing Juice

Prep Time: 10 minutes
Portion Size: 2

Ingredients:
- 2 apples, cored and chopped
- 2 carrots, peeled and chopped
- 1 tablespoon lemon juice

Instructions:
1. Combine apples and carrots in a juicer or blender.
2. Blend until smooth.
3. Stir in lemon juice and pour into glasses.

Nutritional Information (per serving):
- Calories: 90
- Carbohydrates: 22g
- Protein: 1g
- Fat: 0g

Potassium-Rich Banana and Spinach Smoothie

Prep Time: 5 minutes
Portion Size: 2

Ingredients:
- 1 banana
- 1 cup fresh spinach
- 1 cup almond milk
- 1 tablespoon chia seeds

Instructions:
1. Combine all ingredients in a blender.
2. Blend until smooth.
3. Pour into glasses and serve immediately.

Nutritional Information (per serving):
- Calories: 120
- Carbohydrates: 24g
- Protein: 3g
- Fat: 2g

Anti-Oxidant Blueberry and Beet Juice

Prep Time: 10 minutes
Portion Size: 2

Ingredients:
- 1 cup blueberries
- 1 beet, peeled and chopped
- 1 apple, cored and chopped
- 1 tablespoon lemon juice

Instructions:
1. Combine blueberries, beet, and apple in a juicer or blender.
2. Blend until smooth.
3. Stir in lemon juice and pour into glasses.

Nutritional Information (per serving):
- Calories: 100
- Carbohydrates: 24g
- Protein: 1g
- Fat: 0g

Hydrating Coconut Water and Aloe Vera Drink

Prep Time: 5 minutes
Portion Size: 2

Ingredients:
- 2 cups coconut water
- 1/4 cup aloe vera juice
- 1 tablespoon lemon juice

Instructions:
1. Combine coconut water, aloe vera juice, and lemon juice in a pitcher.
2. Stir well and pour into glasses.

3. Serve chilled.

Nutritional Information (per serving):
- Calories: 30
- Carbohydrates: 7g
- Protein: 0g
- Fat: 0g

Detox Green Apple and Kale Smoothie

Prep Time: 5 minutes
Portion Size: 2

Ingredients:
- 1 green apple, cored and chopped
- 1 cup fresh kale
- 1 cup water
- 1 tablespoon honey (optional)

Instructions:
1. Combine all ingredients in a blender.
2. Blend until smooth.
3. Pour into glasses and serve immediately.

Nutritional Information (per serving):
- Calories: 80
- Carbohydrates: 20g
- Protein: 1g
- Fat: 0g

Anti-Inflammatory Pineapple and Turmeric Juice

Prep Time: 5 minutes
Portion Size: 2

Ingredients:
- 1 cup pineapple chunks (fresh or frozen)
- 1 teaspoon turmeric powder
- 1 cup water
- 1 tablespoon honey (optional)

Instructions:
1. Combine all ingredients in a blender.
2. Blend until smooth.
3. Pour into glasses and serve immediately.

Nutritional Information (per serving):
- Calories: 70
- Carbohydrates: 18g
- Protein: 0g
- Fat: 0g

Lemon and Ginger Detox Water

Prep Time: 5 minutes
Portion Size: 2

Ingredients:
- 1 lemon, sliced
- 1 teaspoon fresh grated ginger
- 4 cups water

Instructions:
1. Combine lemon slices and grated ginger in a pitcher.
2. Add water and stir well.
3. Refrigerate for at least 1 hour before serving.

Nutritional Information (per serving):
- Calories: 5
- Carbohydrates: 1g
- Protein: 0g
- Fat: 0g

Healing Aloe Vera and Cucumber Juice

Prep Time: 10 minutes
Portion Size: 2

Ingredients:
- 1/4 cup aloe vera juice
- 1 cucumber, chopped
- 1 apple, cored and chopped

- 1 tablespoon lemon juice

Instructions:
1. Combine aloe vera juice, cucumber, and apple in a blender.
2. Blend until smooth.
3. Stir in lemon juice and pour into glasses.

Nutritional Information (per serving):
- Calories: 50
- Carbohydrates: 13g
- Protein: 1g
- Fat: 0g

Anti-Oxidant Rich Pomegranate and Beet Juice

Prep Time: 10 minutes
Portion Size: 2

Ingredients:
- 1 cup pomegranate seeds
- 1 beet, peeled and chopped
- 1 apple, cored and chopped
- 1 tablespoon lemon juice

Instructions:
1. Combine pomegranate seeds, beet, and apple in a juicer or blender.
2. Blend until smooth.
3. Stir in lemon juice and pour into glasses.

Nutritional Information (per serving):
- Calories: 100
- Carbohydrates: 24g
- Protein: 1g
- Fat: 0g

Cleansing Celery and Apple Smoothie

Prep Time: 5 minutes
Portion Size: 2

Ingredients:

- 2 celery stalks
- 1 apple, cored and chopped
- 1 cup water
- 1 tablespoon honey (optional)

Instructions:
1. Combine all ingredients in a blender.
2. Blend until smooth.
3. Pour into glasses and serve immediately.

Nutritional Information (per serving):
- Calories: 50
- Carbohydrates: 13g
- Protein: 1g
- Fat: 0g

Hydration Boosting Coconut and Lime Drink

Prep Time: 5 minutes
Portion Size: 2

Ingredients:
- 2 cups coconut water
- Juice of 1 lime
- 1 tablespoon honey (optional)

Instructions:
1. Combine coconut water, lime juice, and honey in a pitcher.
2. Stir well and pour into glasses.
3. Serve chilled.

Nutritional Information (per serving):
- Calories: 30
- Carbohydrates: 7g
- Protein: 0g
- Fat: 0g

Anti-Inflammatory Ginger and Lemon Tea

Prep Time: 5 minutes
Portion Size: 2

Ingredients:
- 2 cups water
- 1 teaspoon fresh grated ginger
- Juice of 1 lemon
- 1 tablespoon honey (optional)

Instructions:
1. Bring water to a boil.
2. Add grated ginger, reduce heat, and simmer for 5 minutes.
3. Strain into cups and add lemon juice and honey if desired.

Nutritional Information (per serving):
- Calories: 15
- Carbohydrates: 4g
- Protein: 0g
- Fat: 0g

Detoxifying Green Tea and Cranberry Drink

Prep Time: 5 minutes
Portion Size: 2

Ingredients:
- 1 cup brewed green tea, cooled
- 1 cup unsweetened cranberry juice
- 1 tablespoon honey (optional)

Instructions:
1. Brew green tea and let it cool.
2. Mix green tea with cranberry juice in a pitcher.
3. Stir well and add honey if desired.
4. Pour into glasses and serve chilled.

Nutritional Information (per serving):
- Calories: 30
- Carbohydrates: 8g
- Protein: 0g
- Fat: 0g

Kidney Health Watermelon and Cucumber Juice

Prep Time: 10 minutes
Portion Size: 2

Ingredients:
- 2 cups diced watermelon
- 1 cucumber, chopped
- Juice of 1 lime
- 1 tablespoon honey (optional)

Instructions:
1. Blend watermelon, cucumber, and lime juice until smooth.
2. Strain the mixture to remove pulp, if desired.
3. Pour into glasses and serve chilled.

Nutritional Information (per serving):
- Calories: 50
- Carbohydrates: 13g
- Protein: 1g
- Fat: 0g

Dietary Tips for Kidney Health

1. **Stay Hydrated:** Drink plenty of water throughout the day to help your kidneys flush out toxins and waste products. Aim for at least 8 glasses of water daily, unless advised otherwise by your healthcare provider.

2. **Limit Sodium Intake:** Excessive salt can increase blood pressure and strain the kidneys. Choose fresh, unprocessed foods and avoid adding extra salt to your meals. Read food labels to monitor sodium levels.

3. **Eat a Balanced Diet:** Incorporate a variety of fruits, vegetables, whole grains, and lean proteins into your diet. This provides essential nutrients that support overall kidney health.

4. **Choose Low-Phosphorus Foods:** High levels of phosphorus can be harmful to your kidneys. Limit foods like dairy products, processed meats, and soda. Opt for fresh fruits and vegetables, and grains like rice and oats.

5. **Monitor Protein Intake:** While protein is essential for health, too much can put extra strain on your kidneys. Balance your protein sources by including plant-based options like beans, lentils, and nuts.

6. **Limit Potassium-Rich Foods:** If you have kidney disease, your doctor may advise you to limit potassium-rich foods such as bananas, oranges, potatoes, and spinach. Follow your healthcare provider's recommendations regarding potassium intake.

7. **Avoid Processed Foods:** Processed and packaged foods often contain high levels of sodium, phosphorus, and unhealthy fats. Choose fresh, whole foods to support kidney health.

8. **Reduce Sugar Intake:** Excess sugar can lead to weight gain and increased risk of diabetes, which can harm your kidneys. Limit sugary drinks, candies, and desserts.

9. **Eat Healthy Fats:** Include sources of healthy fats in your diet, such as avocados, olive oil, and fatty fish like salmon. These fats can help reduce inflammation and support overall health.

10. **Avoid Excessive Alcohol and Caffeine:** Both alcohol and caffeine can dehydrate the body and put stress on the kidneys. Consume these beverages in moderation.

Natural Recipes for Liver Health

Green Detox Smoothie with Kale and Lemon

Prep Time: 10 minutes
Cook Time: 0 minutes
Portion Size: 2

Ingredients:
- 2 cups kale leaves, chopped
- 1 banana
- 1/2 lemon, juiced
- 1 apple, cored and chopped
- 1 cup coconut water
- 1 tablespoon chia seeds
- 1 teaspoon honey (optional)

Instructions:
1. Combine all ingredients in a blender.
2. Blend until smooth.
3. Pour into glasses and serve immediately.

Nutritional Information (per serving):
- Calories: 120
- Carbohydrates: 30g
- Protein: 3g
- Fat: 1g

Turmeric and Ginger Liver Cleanse Tea

Prep Time: 5 minutes
Cook Time: 10 minutes
Portion Size: 2

Ingredients:
- 2 cups water
- 1 teaspoon turmeric powder
- 1-inch piece of fresh ginger, grated
- 1 tablespoon lemon juice
- 1 teaspoon honey (optional)

Instructions:
1. Boil water in a pot.
2. Add turmeric and ginger.

3. Simmer for 10 minutes.
4. Strain into cups.
5. Stir in lemon juice and honey.

Nutritional Information (per serving):
- Calories: 15
- Carbohydrates: 4g
- Protein: 0g
- Fat: 0g

Beetroot and Carrot Liver Tonic Juice

Prep Time: 10 minutes
Cook Time: 0 minutes
Portion Size: 2

Ingredients:
- 2 medium beetroots, peeled and chopped
- 3 carrots, peeled and chopped
- 1 apple, cored and chopped
- 1-inch piece of ginger, peeled
- 1/2 lemon, juiced

Instructions:
1. Process all ingredients through a juicer.
2. Stir well and serve immediately.

Nutritional Information (per serving):
- Calories: 110
- Carbohydrates: 25g
- Protein: 2g
- Fat: 0g

Avocado and Spinach Salad with Apple Cider Vinaigrette

Prep Time: 15 minutes
Cook Time: 0 minutes
Portion Size: 2

Ingredients:
- 2 cups spinach leaves

- 1 avocado, sliced
- 1/2 cucumber, sliced
- 1/4 red onion, thinly sliced
- 1/4 cup cherry tomatoes, halved
- 2 tablespoons apple cider vinegar
- 1 tablespoon olive oil
- Salt and pepper to taste

Instructions:
1. Combine spinach, avocado, cucumber, onion, and tomatoes in a bowl.
2. Whisk together vinegar, olive oil, salt, and pepper.
3. Drizzle dressing over salad and toss gently.

Nutritional Information (per serving):
- Calories: 220
- Carbohydrates: 12g
- Protein: 3g
- Fat: 20g

Cilantro and Lime Liver Detox Soup

Prep Time: 10 minutes
Cook Time: 20 minutes
Portion Size: 2

Ingredients:
- 1 tablespoon olive oil
- 1 onion, chopped
- 2 cloves garlic, minced
- 1 zucchini, chopped
- 1 cup vegetable broth
- 1/2 cup fresh cilantro, chopped
- Juice of 1 lime
- Salt and pepper to taste

Instructions:
1. Heat olive oil in a pot over medium heat.
2. Sauté onion and garlic until fragrant.
3. Add zucchini and cook for 5 minutes.
4. Pour in vegetable broth and bring to a boil.
5. Reduce heat and simmer for 10 minutes.
6. Stir in cilantro and lime juice.
7. Season with salt and pepper.

Nutritional Information (per serving):
- Calories: 100
- Carbohydrates: 10g
- Protein: 2g
- Fat: 7g

Dandelion Greens and Quinoa Liver Power Bowl

Prep Time: 15 minutes
Cook Time: 15 minutes
Portion Size: 2

Ingredients:
- 1 cup cooked quinoa
- 2 cups dandelion greens, chopped
- 1/2 cup cherry tomatoes, halved
- 1/2 avocado, sliced
- 1/4 cup red onion, thinly sliced
- 1 tablespoon olive oil
- Juice of 1 lemon
- Salt and pepper to taste

Instructions:
1. Divide quinoa between two bowls.
2. Top with dandelion greens, tomatoes, avocado, and onion.
3. Drizzle with olive oil and lemon juice.
4. Season with salt and pepper.

Nutritional Information (per serving):
- Calories: 250
- Carbohydrates: 30g
- Protein: 7g
- Fat: 12g

Lemon and Mint Liver Flush Water

Prep Time: 5 minutes
Cook Time: 0 minutes
Portion Size: 2

Ingredients:
- 1 lemon, sliced
- 1/2 cucumber, sliced
- 10 fresh mint leaves
- 4 cups water

Instructions:
1. Combine all ingredients in a pitcher.
2. Refrigerate for at least 1 hour.
3. Serve chilled.

Nutritional Information (per serving):
- Calories: 5
- Carbohydrates: 1g
- Protein: 0g
- Fat: 0g

Broccoli and Turmeric Stir-Fry with Brown Rice

Prep Time: 10 minutes
Cook Time: 15 minutes
Portion Size: 2

Ingredients:
- 1 tablespoon olive oil
- 2 cups broccoli florets
- 1 bell pepper, sliced
- 1 teaspoon turmeric powder
- 1 cup cooked brown rice
- 1 tablespoon soy sauce
- Salt and pepper to taste

Instructions:
1. Heat olive oil in a skillet over medium heat.
2. Add broccoli and bell pepper; cook for 5 minutes.
3. Sprinkle turmeric powder and stir well.
4. Add brown rice and soy sauce.
5. Cook for another 5 minutes.
6. Season with salt and pepper.

Nutritional Information (per serving):
- Calories: 250
- Carbohydrates: 40g
- Protein: 6g

- Fat: 8g

Garlic and Olive Oil Liver Detox Dressing

Prep Time: 5 minutes
Cook Time: 0 minutes
Portion Size: 2

Ingredients:
- 1/4 cup olive oil
- 2 cloves garlic, minced
- 2 tablespoons lemon juice
- 1 tablespoon apple cider vinegar
- Salt and pepper to taste

Instructions:
1. Whisk together all ingredients in a bowl.
2. Drizzle over salads or vegetables.

Nutritional Information (per serving):
- Calories: 100
- Carbohydrates: 2g
- Protein: 0g
- Fat: 10g

Apple, Cucumber, and Celery Liver Cleanse Juice

Prep Time: 10 minutes
Cook Time: 0 minutes
Portion Size: 2

Ingredients:
- 2 apples, cored and chopped
- 1 cucumber, chopped
- 3 celery stalks, chopped
- 1/2 lemon, juiced

Instructions:
1. Process all ingredients through a juicer.
2. Stir well and serve immediately.

Nutritional Information (per serving):
- Calories: 90
- Carbohydrates: 22g
- Protein: 1g
- Fat: 0g

Spirulina and Pineapple Liver Boost Smoothie

Prep Time: 10 minutes
Cook Time: 0 minutes
Portion Size: 2

Ingredients:
- 1 cup fresh pineapple, chopped
- 1 banana
- 1 teaspoon spirulina powder
- 1 cup coconut water
- 1 tablespoon chia seeds

Instructions:
1. Combine all ingredients in a blender.
2. Blend until smooth.
3. Pour into glasses and serve immediately.

Nutritional Information (per serving):
- Calories: 150
- Carbohydrates: 35g
- Protein: 3g
- Fat: 2g

Asparagus and Lemon Zest Detox Salad

Prep Time: 10 minutes
Cook Time: 5 minutes
Portion Size: 2

Ingredients:
- 1 bunch asparagus, trimmed

- 1 tablespoon olive oil
- Zest of 1 lemon
- Juice of 1/2 lemon
- Salt and pepper to taste

Instructions:
1. Blanch asparagus in boiling water for 2-3 minutes.
2. Drain and rinse under cold water.
3. Toss with olive oil, lemon zest, and lemon juice.
4. Season with salt and pepper.

Nutritional Information (per serving):
- Calories: 70
- Carbohydrates: 5g
- Protein: 2g
- Fat: 5g

Chia Seed and Blueberry Liver Rejuvenation Pudding

Prep Time: 10 minutes
Cook Time: 0 minutes
Portion Size: 2

Ingredients:
- 1/4 cup chia seeds
- 1 cup almond milk
- 1/2 cup blueberries
- 1 teaspoon honey (optional)

Instructions:
1. Combine chia seeds and almond milk in a bowl.
2. Stir well and refrigerate for at least 2 hours.
3. Top with blueberries and honey before serving.

Nutritional Information (per serving):
- Calories: 150
- Carbohydrates: 20g
- Protein: 4g
- Fat: 7g

Bitter Melon and Ginger Healing Broth

Prep Time: 10 minutes
Cook Time: 20 minutes
Portion Size: 2

Ingredients:
- 1 tablespoon olive oil
- 1 onion, chopped
- 2 cloves garlic, minced
- 1 bitter melon, sliced
- 1-inch piece of ginger, grated
- 4 cups vegetable broth
- Salt and pepper to taste

Instructions:
1. Heat olive oil in a pot over medium heat.
2. Sauté onion and garlic until softened.
3. Add bitter melon and ginger; cook for 5 minutes.
4. Pour in vegetable broth and bring to a boil.
5. Reduce heat and simmer for 10 minutes.
6. Season with salt and pepper.

Nutritional Information (per serving):
- Calories: 80
- Carbohydrates: 10g
- Protein: 2g
- Fat: 4g

Red Cabbage and Beet Liver Support Slaw

Prep Time: 15 minutes
Cook Time: 0 minutes
Portion Size: 2

Ingredients:
- 1 cup shredded red cabbage
- 1 beet, peeled and grated
- 1 carrot, peeled and grated
- 2 tablespoons apple cider vinegar
- 1 tablespoon olive oil
- Salt and pepper to taste

Instructions:
1. Combine red cabbage, beet, and carrot in a bowl.
2. Whisk together apple cider vinegar and olive oil.

3. Pour dressing over the slaw and toss to combine.
4. Season with salt and pepper.

Nutritional Information (per serving):
- Calories: 90
- Carbohydrates: 12g
- Protein: 2g
- Fat: 5g

Artichoke and Lemon Roasted Veggie Medley

Prep Time: 10 minutes
Cook Time: 25 minutes
Portion Size: 2

Ingredients:
- 1 cup artichoke hearts, quartered
- 1 zucchini, chopped
- 1 bell pepper, chopped
- 1 tablespoon olive oil
- Juice of 1 lemon
- Salt and pepper to taste

Instructions:
1. Preheat oven to 400°F (200°C).
2. Toss artichoke hearts, zucchini, and bell pepper with olive oil and lemon juice.
3. Spread on a baking sheet.
4. Roast for 20-25 minutes, until vegetables are tender.
5. Season with salt and pepper.

Nutritional Information (per serving):
- Calories: 150
- Carbohydrates: 15g
- Protein: 3g
- Fat: 8g

Flaxseed and Papaya Liver Revitalizing Smoothie

Prep Time: 10 minutes
Cook Time: 0 minutes

Portion Size: 2

Ingredients:
- 1 cup fresh papaya, chopped
- 1 banana
- 1 tablespoon flaxseed
- 1 cup almond milk
- 1 teaspoon honey (optional)

Instructions:
1. Combine all ingredients in a blender.
2. Blend until smooth.
3. Pour into glasses and serve immediately.

Nutritional Information (per serving):
- Calories: 140
- Carbohydrates: 28g
- Protein: 3g
- Fat: 3g

Wheatgrass and Orange Morning Detox Juice

Prep Time: 10 minutes
Cook Time: 0 minutes
Portion Size: 2

Ingredients:
- 2 oranges, peeled and chopped
- 1 cup wheatgrass
- 1/2 lemon, juiced
- 1-inch piece of ginger, peeled

Instructions:
1. Process all ingredients through a juicer.
2. Stir well and serve immediately.

Nutritional Information (per serving):
- Calories: 70
- Carbohydrates: 17g
- Protein: 1g
- Fat: 0g

Parsley and Lemon Liver Cleansing Soup

Prep Time: 10 minutes
Cook Time: 20 minutes
Portion Size: 2

Ingredients:
- 1 tablespoon olive oil
- 1 onion, chopped
- 2 cloves garlic, minced
- 1 bunch parsley, chopped
- 4 cups vegetable broth
- Juice of 1 lemon
- Salt and pepper to taste

Instructions:
1. Heat olive oil in a pot over medium heat.
2. Sauté onion and garlic until softened.
3. Add parsley and cook for 5 minutes.
4. Pour in vegetable broth and bring to a boil.
5. Reduce heat and simmer for 15 minutes.
6. Stir in lemon juice.
7. Season with salt and pepper.

Nutritional Information (per serving):
- Calories: 90
- Carbohydrates: 10g
- Protein: 2g
- Fat: 5g

Green Apple and Kale Liver Purification Salad

Prep Time: 15 minutes
Cook Time: 0 minutes
Portion Size: 2

Ingredients:
- 2 cups kale leaves, chopped
- 1 green apple, cored and sliced
- 1/4 cup walnuts, chopped
- 2 tablespoons apple cider vinegar
- 1 tablespoon olive oil

- Salt and pepper to taste

Instructions:

1. Combine kale, apple, and walnuts in a bowl.
2. Whisk together apple cider vinegar and olive oil.
3. Drizzle dressing over salad and toss to combine.
4. Season with salt and pepper.

Nutritional Information (per serving):

- Calories: 180
- Carbohydrates: 15g
- Protein: 3g
- Fat: 12g

Tips for a Detoxifying Diet

1. **Increase Water Intake**: Drink at least 8-10 glasses of water daily to help flush toxins from your body. Consider starting your day with a glass of warm water with lemon.
2. **Eat More Vegetables**: Aim to fill half your plate with vegetables at each meal. Cruciferous vegetables like broccoli, cauliflower, and Brussels sprouts are especially beneficial for liver detoxification.
3. **Incorporate Leafy Greens**: Include leafy greens such as kale, spinach, and dandelion greens in your diet. These are rich in chlorophyll, which helps cleanse the blood and support liver function.
4. **Choose Organic Foods**: Whenever possible, choose organic fruits and vegetables to reduce exposure to pesticides and other harmful chemicals.
5. **Limit Processed Foods**: Avoid processed and packaged foods that are high in additives, preservatives, and unhealthy fats. Focus on whole, unprocessed foods.
6. **Consume Healthy Fats**: Include sources of healthy fats in your diet such as avocados, nuts, seeds, and olive oil. These fats support cell function and help absorb fat-soluble vitamins.
7. **Reduce Sugar Intake**: Minimize consumption of refined sugars and artificial sweeteners. Instead, opt for natural sweeteners like honey, maple syrup, or fruits.
8. **Incorporate Probiotics**: Eat fermented foods such as yogurt, kefir, sauerkraut, and kimchi to support a healthy gut microbiome, which is crucial for detoxification.
9. **Drink Herbal Teas**: Herbal teas such as dandelion, milk thistle, and green tea have detoxifying properties and can support liver health.
10. **Exercise Regularly**: Engage in regular physical activity to help improve circulation and promote the elimination of toxins through sweat.
11. **Get Adequate Sleep**: Aim for 7-9 hours of sleep per night. Quality sleep is essential for the body's natural detoxification processes.
12. **Avoid Alcohol and Caffeine**: Limit or eliminate alcohol and caffeine consumption, as these can place additional stress on the liver.
13. **Eat Fiber-Rich Foods**: Incorporate plenty of fiber-rich foods such as fruits, vegetables, whole grains, and legumes to support digestive health and regular bowel movements.
14. **Practice Mindful Eating**: Eat slowly and mindfully, paying attention to hunger and fullness cues. This helps improve digestion and absorption of nutrients.
15. **Plan Balanced Meals**: Aim for balanced meals that include a variety of nutrients. Ensure each meal contains a mix of protein, healthy fats, and complex carbohydrates.
16. **Stay Active**: Incorporate activities like yoga, stretching, and walking into your daily routine to support overall well-being and reduce stress.
17. **Avoid Environmental Toxins**: Reduce exposure to environmental toxins by using natural cleaning products, avoiding plastic containers, and choosing personal care products with fewer chemicals.

Natural Recipes for Bone Health

Calcium-Rich Kale and Almond Smoothie

Prep Time: 10 minutes
Cook Time: 0 minutes
Portion Size: 2

Ingredients:
- 2 cups kale leaves, chopped
- 1 banana
- 1/2 cup almond milk
- 1/4 cup almonds, soaked
- 1 tablespoon chia seeds
- 1 teaspoon honey (optional)

Instructions:
1. Combine all ingredients in a blender.
2. Blend until smooth.
3. Pour into glasses and serve immediately.

Nutritional Information (per serving):
- Calories: 180
- Carbohydrates: 22g
- Protein: 5g
- Fat: 9g

Bone-Strengthening Broccoli and Salmon Salad

Prep Time: 15 minutes
Cook Time: 10 minutes
Portion Size: 2

Ingredients:
- 2 cups broccoli florets, steamed
- 1 salmon fillet, cooked and flaked
- 1/2 cup cherry tomatoes, halved
- 1/4 red onion, thinly sliced
- 1 tablespoon olive oil
- Juice of 1 lemon
- Salt and pepper to taste

Instructions:
1. Combine broccoli, salmon, cherry tomatoes, and red onion in a bowl.

2. Drizzle with olive oil and lemon juice.
3. Toss gently to combine.
4. Season with salt and pepper.

Nutritional Information (per serving):
- Calories: 250
- Carbohydrates: 10g
- Protein: 22g
- Fat: 14g

Vitamin D-Fortified Mushroom and Spinach Omelette

Prep Time: 10 minutes
Cook Time: 10 minutes
Portion Size: 2

Ingredients:
- 4 eggs
- 1/2 cup mushrooms, sliced
- 1 cup spinach leaves
- 1 tablespoon olive oil
- Salt and pepper to taste

Instructions:
1. Whisk eggs in a bowl.
2. Heat olive oil in a skillet over medium heat.
3. Add mushrooms and cook until softened.
4. Add spinach and cook until wilted.
5. Pour eggs into the skillet.
6. Cook until the eggs are set.
7. Season with salt and pepper.

Nutritional Information (per serving):
- Calories: 200
- Carbohydrates: 4g
- Protein: 14g
- Fat: 15g

Magnesium-Boosting Avocado and Quinoa Bowl

Prep Time: 15 minutes
Cook Time: 15 minutes
Portion Size: 2

Ingredients:
- 1 cup cooked quinoa
- 1 avocado, sliced
- 1/2 cup cherry tomatoes, halved
- 1/4 cup red onion, thinly sliced
- 2 tablespoons pumpkin seeds
- 1 tablespoon olive oil
- Juice of 1 lime
- Salt and pepper to taste

Instructions:
1. Divide quinoa between two bowls.
2. Top with avocado, cherry tomatoes, red onion, and pumpkin seeds.
3. Drizzle with olive oil and lime juice.
4. Season with salt and pepper.

Nutritional Information (per serving):
- Calories: 300
- Carbohydrates: 34g
- Protein: 7g
- Fat: 17g

Bone-Building Sesame and Chickpea Hummus

Prep Time: 10 minutes
Cook Time: 0 minutes
Portion Size: 2

Ingredients:
- 1 can chickpeas, rinsed and drained
- 2 tablespoons tahini
- 1 clove garlic
- Juice of 1 lemon
- 1 tablespoon olive oil
- 1/4 cup water
- Salt and pepper to taste

Instructions:
1. Combine all ingredients in a food processor.
2. Blend until smooth.

3. Serve with vegetables or crackers.

Nutritional Information (per serving):
- Calories: 180
- Carbohydrates: 18g
- Protein: 6g
- Fat: 9g

Calcium-Loaded Orange and Fig Breakfast Parfait

Prep Time: 10 minutes
Cook Time: 0 minutes
Portion Size: 2

Ingredients:
- 1 cup Greek yogurt
- 1 orange, peeled and segmented
- 4 dried figs, chopped
- 2 tablespoons granola
- 1 tablespoon honey (optional)

Instructions:
1. Layer yogurt, orange segments, and figs in two glasses.
2. Top with granola and honey.
3. Serve immediately.

Nutritional Information (per serving):
- Calories: 250
- Carbohydrates: 38g
- Protein: 12g
- Fat: 6g

Anti-Inflammatory Turmeric and Ginger Bone Broth

Prep Time: 10 minutes
Cook Time: 45 minutes
Portion Size: 2

Ingredients:
- 4 cups chicken bone broth
- 1 teaspoon turmeric powder

- 1-inch piece of ginger, sliced
- 1 clove garlic, minced
- Juice of 1 lemon
- Salt and pepper to taste

Instructions:
1. Combine all ingredients in a pot.
2. Bring to a boil, then reduce heat and simmer for 45 minutes.
3. Strain and serve hot.

Nutritional Information (per serving):
- Calories: 60
- Carbohydrates: 2g
- Protein: 6g
- Fat: 2g

Bone-Healthy Sardine and Arugula Salad

Prep Time: 10 minutes
Cook Time: 0 minutes
Portion Size: 2

Ingredients:
- 2 cups arugula
- 1 can sardines, drained and flaked
- 1/2 cucumber, sliced
- 1/4 red onion, thinly sliced
- 1 tablespoon olive oil
- Juice of 1 lemon
- Salt and pepper to taste

Instructions:
1. Combine arugula, sardines, cucumber, and red onion in a bowl.
2. Drizzle with olive oil and lemon juice.
3. Toss gently to combine.
4. Season with salt and pepper.

Nutritional Information (per serving):
- Calories: 220
- Carbohydrates: 6g
- Protein: 15g
- Fat: 15g

Collagen-Rich Chicken and Vegetable Soup

Prep Time: 15 minutes
Cook Time: 30 minutes
Portion Size: 2

Ingredients:
- 1 tablespoon olive oil
- 1 onion, chopped
- 2 cloves garlic, minced
- 2 carrots, sliced
- 1 cup cooked chicken, shredded
- 4 cups chicken broth
- 1 cup spinach leaves
- Salt and pepper to taste

Instructions:
1. Heat olive oil in a pot over medium heat.
2. Sauté onion and garlic until softened.
3. Add carrots and cook for 5 minutes.
4. Pour in chicken broth and bring to a boil.
5. Reduce heat and simmer for 20 minutes.
6. Add shredded chicken and spinach.
7. Cook for another 5 minutes.
8. Season with salt and pepper.

Nutritional Information (per serving):
- Calories: 200
- Carbohydrates: 12g
- Protein: 20g
- Fat: 8g

Bone-Bolstering Tahini and Carrot Slaw

Prep Time: 10 minutes
Cook Time: 0 minutes
Portion Size: 2

Ingredients:

- 2 cups shredded carrots
- 1/4 cup tahini
- 1 tablespoon apple cider vinegar
- 1 tablespoon olive oil
- 1 tablespoon lemon juice
- Salt and pepper to taste

Instructions:
1. Combine shredded carrots in a bowl.
2. Whisk together tahini, apple cider vinegar, olive oil, and lemon juice.
3. Pour dressing over carrots and toss to combine.
4. Season with salt and pepper.

Nutritional Information (per serving):
- Calories: 150
- Carbohydrates: 12g
- Protein: 3g
- Fat: 10g

Calcium-Packed Almond and Berry Overnight Oats

Prep Time: 10 minutes
Cook Time: 0 minutes
Portion Size: 2

Ingredients:
- 1 cup rolled oats
- 1 cup almond milk
- 1/4 cup almonds, chopped
- 1/2 cup mixed berries
- 1 tablespoon chia seeds
- 1 teaspoon honey (optional)

Instructions:
1. Combine all ingredients in a bowl.
2. Mix well and refrigerate overnight.
3. Serve chilled in the morning.

Nutritional Information (per serving):
- Calories: 250
- Carbohydrates: 38g
- Protein: 7g
- Fat: 8g

Bone-Supporting Lentil and Spinach Stew

Prep Time: 15 minutes
Cook Time: 30 minutes
Portion Size: 2

Ingredients:
- 1 tablespoon olive oil
- 1 onion, chopped
- 2 cloves garlic, minced
- 1 cup lentils, rinsed
- 4 cups vegetable broth
- 2 cups spinach leaves
- 1 teaspoon cumin
- Salt and pepper to taste

Instructions:
1. Heat olive oil in a pot over medium heat.
2. Sauté onion and garlic until softened.
3. Add lentils and vegetable broth.
4. Bring to a boil, then reduce heat and simmer for 25 minutes.
5. Stir in spinach and cumin.
6. Cook for another 5 minutes.
7. Season with salt and pepper.

Nutritional Information (per serving):
- Calories: 250
- Carbohydrates: 35g
- Protein: 12g
- Fat: 6g

Magnesium-Enriched Sweet Potato and Black Bean Tacos

Prep Time: 15 minutes
Cook Time: 20 minutes
Portion Size: 2

Ingredients:
- 1 large sweet potato, peeled and diced
- 1 can black beans, rinsed and drained
- 1 tablespoon olive oil
- 1/2 teaspoon cumin

- 1/2 teaspoon paprika
- Salt and pepper to taste
- 4 small corn tortillas
- 1/4 cup fresh cilantro, chopped
- 1/2 avocado, sliced

Instructions:

1. Preheat oven to 400°F (200°C).
2. Toss sweet potato with olive oil, cumin, paprika, salt, and pepper.
3. Spread on a baking sheet and roast for 20 minutes, until tender.
4. Warm tortillas in a skillet over medium heat.
5. Fill tortillas with sweet potato and black beans.
6. Top with cilantro and avocado slices.

Nutritional Information (per serving):
- Calories: 350
- Carbohydrates: 60g
- Protein: 10g
- Fat: 10g

Omega-3 Rich Walnut and Blueberry Salad

Prep Time: 10 minutes
Cook Time: 0 minutes
Portion Size: 2

Ingredients:
- 4 cups mixed greens
- 1/2 cup blueberries
- 1/4 cup walnuts, chopped
- 1/4 cup feta cheese, crumbled
- 2 tablespoons olive oil
- 1 tablespoon balsamic vinegar
- Salt and pepper to taste

Instructions:

1. Combine mixed greens, blueberries, walnuts, and feta in a bowl.
2. Whisk together olive oil and balsamic vinegar.
3. Drizzle dressing over salad and toss to combine.
4. Season with salt and pepper.

Nutritional Information (per serving):
- Calories: 220
- Carbohydrates: 15g

- Protein: 6g
- Fat: 16g

Bone-Nourishing Butternut Squash and Kale Risotto

Prep Time: 15 minutes
Cook Time: 30 minutes
Portion Size: 2

Ingredients:
- 1 tablespoon olive oil
- 1 onion, chopped
- 2 cloves garlic, minced
- 1 cup butternut squash, diced
- 1 cup arborio rice
- 4 cups vegetable broth, warmed
- 2 cups kale, chopped
- 1/4 cup grated Parmesan cheese
- Salt and pepper to taste

Instructions:
1. Heat olive oil in a pot over medium heat.
2. Sauté onion and garlic until softened.
3. Add butternut squash and cook for 5 minutes.
4. Stir in arborio rice and cook for 2 minutes.
5. Add vegetable broth, one cup at a time, stirring constantly until absorbed.
6. Stir in kale and cook until wilted.
7. Remove from heat and stir in Parmesan cheese.
8. Season with salt and pepper.

Nutritional Information (per serving):
- Calories: 350
- Carbohydrates: 65g
- Protein: 10g
- Fat: 8g

Calcium-Fortified Chia and Coconut Pudding

Prep Time: 10 minutes
Cook Time: 0 minutes

Portion Size: 2

Ingredients:
- 1/4 cup chia seeds
- 1 cup coconut milk
- 1 tablespoon honey
- 1/2 teaspoon vanilla extract
- 1/4 cup shredded coconut
- 1/2 cup fresh berries

Instructions:
1. Combine chia seeds, coconut milk, honey, and vanilla extract in a bowl.
2. Mix well and refrigerate for at least 2 hours.
3. Top with shredded coconut and fresh berries before serving.

Nutritional Information (per serving):
- Calories: 250
- Carbohydrates: 20g
- Protein: 4g
- Fat: 18g

Bone-Beneficial Edamame and Sesame Stir-Fry

Prep Time: 10 minutes
Cook Time: 10 minutes
Portion Size: 2

Ingredients:
- 1 tablespoon sesame oil
- 2 cups edamame, shelled
- 1 bell pepper, sliced
- 1 carrot, julienned
- 2 tablespoons soy sauce
- 1 tablespoon sesame seeds
- 2 green onions, chopped

Instructions:
1. Heat sesame oil in a skillet over medium heat.
2. Add edamame, bell pepper, and carrot; cook for 5 minutes.
3. Stir in soy sauce and cook for another 3 minutes.
4. Sprinkle with sesame seeds and green onions.
5. Serve hot.

Nutritional Information (per serving):
- Calories: 220

- Carbohydrates: 18g
- Protein: 12g
- Fat: 12g

Phosphorus-Packed Pumpkin Seed and Avocado Toast

Prep Time: 10 minutes
Cook Time: 0 minutes
Portion Size: 2

Ingredients:
- 2 slices whole grain bread, toasted
- 1 avocado, mashed
- 2 tablespoons pumpkin seeds
- Juice of 1/2 lemon
- Salt and pepper to taste

Instructions:
1. Spread mashed avocado on toasted bread.
2. Sprinkle with pumpkin seeds.
3. Drizzle with lemon juice.
4. Season with salt and pepper.

Nutritional Information (per serving):
- Calories: 250
- Carbohydrates: 30g
- Protein: 6g
- Fat: 15g

Bone-Strengthening Greek Yogurt and Almond Smoothie Bowl

Prep Time: 10 minutes
Cook Time: 0 minutes
Portion Size: 2

Ingredients:
- 1 cup Greek yogurt
- 1 banana, sliced
- 1/4 cup almonds, chopped
- 1/4 cup granola
- 1 tablespoon honey

- 1/2 cup mixed berries

Instructions:
1. Divide Greek yogurt between two bowls.
2. Top with banana slices, almonds, granola, and mixed berries.
3. Drizzle with honey.

Nutritional Information (per serving):
- Calories: 300
- Carbohydrates: 40g
- Protein: 12g
- Fat: 12g

Calcium-Enhanced Tofu and Bok Choy Stir-Fry

Prep Time: 10 minutes
Cook Time: 10 minutes
Portion Size: 2

Ingredients:
- 1 tablespoon olive oil
- 1 block firm tofu, cubed
- 2 cups bok choy, chopped
- 1 bell pepper, sliced
- 2 cloves garlic, minced
- 2 tablespoons soy sauce
- 1 tablespoon sesame seeds

Instructions:
1. Heat olive oil in a skillet over medium heat.
2. Add tofu and cook until golden brown.
3. Add bok choy, bell pepper, and garlic; cook for 5 minutes.
4. Stir in soy sauce and cook for another 3 minutes.
5. Sprinkle with sesame seeds before serving.

Nutritional Information (per serving):
- Calories: 250
- Carbohydrates: 10g
- Protein: 18g
- Fat: 15g

Foods Rich in Calcium and Vitamin D

To maintain strong and healthy bones, it's important to include foods rich in calcium and vitamin D in your diet. Here is a list of such foods:

Calcium-Rich Foods
- **Dairy Products**: Milk, cheese, yogurt
- **Leafy Greens**: Kale, spinach, collard greens
- **Fortified Foods**: Calcium-fortified cereals, orange juice, and plant-based milk (almond, soy, rice)
- **Nuts and Seeds**: Almonds, chia seeds, sesame seeds
- **Fish**: Sardines, salmon (with bones)
- **Legumes**: Beans, lentils, chickpeas
- **Tofu**: Especially if prepared with calcium sulfate
- **Broccoli**: A good source of calcium from vegetables

Vitamin D-Rich Foods
- **Fatty Fish**: Salmon, mackerel, tuna
- **Fish Liver Oils**: Cod liver oil
- **Fortified Foods**: Vitamin D-fortified milk, orange juice, cereals, and plant-based milk (almond, soy, rice)
- **Egg Yolks**: A natural source of vitamin D
- **Mushrooms**: Especially those exposed to sunlight or UV light
- **Beef Liver**: Contains small amounts of vitamin D

Natural Recipes for Hormonal Balance for Women and Men

Hormone-Balancing Smoothie with Maca and Berries

Prep Time: 5 minutes
Portion Size: 2

Ingredients:
- 1 cup almond milk
- 1 banana
- 1/2 cup mixed berries (blueberries, strawberries, raspberries)
- 1 tablespoon maca powder
- 1 tablespoon chia seeds
- 1 teaspoon honey
- 1/2 teaspoon cinnamon

Instructions:
1. Combine all ingredients in a blender.
2. Blend until smooth.
3. Pour into glasses and serve immediately.

Nutritional Information (per serving):
- Calories: 150
- Carbohydrates: 30g
- Protein: 4g
- Fat: 3g

Herbal Infusion for Female Hormone Harmony

Prep Time: 10 minutes
Portion Size: 2

Ingredients:
- 2 cups water
- 1 tablespoon dried red clover
- 1 tablespoon dried nettle
- 1 tablespoon dried raspberry leaf
- 1 teaspoon dried chamomile
- 1 teaspoon honey

Instructions:
1. Bring water to a boil.
2. Add herbs to a teapot or heatproof container.
3. Pour boiling water over the herbs.

4. Steep for 10 minutes.
5. Strain the herbs and pour the tea into cups.
6. Add honey to taste.

Nutritional Information (per serving):
- Calories: 20
- Carbohydrates: 5g
- Protein: 0g
- Fat: 0g

Roasted Vegetables with Thyme and Sage for Thyroid Support

Prep Time: 15 minutes
Cook Time: 30 minutes
Portion Size: 2

Ingredients:
- 1 cup chopped carrots
- 1 cup chopped sweet potatoes
- 1 cup chopped Brussels sprouts
- 1 tablespoon olive oil
- 1 teaspoon dried thyme
- 1 teaspoon dried sage
- Salt and pepper to taste

Instructions:
1. Preheat oven to 400°F (200°C).
2. Toss vegetables with olive oil, thyme, sage, salt, and pepper.
3. Spread evenly on a baking sheet.
4. Roast for 30 minutes or until tender and slightly browned.

Nutritional Information (per serving):
- Calories: 180
- Carbohydrates: 30g
- Protein: 3g
- Fat: 7g

Flaxseed and Chia Pudding for Estrogen Balance

Prep Time: 10 minutes (plus overnight refrigeration)

Portion Size: 2

Ingredients:
- 1 cup almond milk
- 2 tablespoons chia seeds
- 2 tablespoons ground flaxseed
- 1 teaspoon vanilla extract
- 1 teaspoon honey
- 1/2 cup fresh berries

Instructions:
1. In a bowl, combine almond milk, chia seeds, flaxseed, vanilla extract, and honey.
2. Stir well to combine.
3. Cover and refrigerate overnight.
4. Stir again before serving, and top with fresh berries.

Nutritional Information (per serving):
- Calories: 200
- Carbohydrates: 20g
- Protein: 6g
- Fat: 10g

Pumpkin Seed Protein Balls for Testosterone Boost

Prep Time: 15 minutes
Portion Size: 2

Ingredients:
- 1/2 cup pumpkin seeds
- 1/2 cup rolled oats
- 1/4 cup almond butter
- 1/4 cup honey
- 1 tablespoon chia seeds
- 1 teaspoon vanilla extract

Instructions:
1. In a food processor, blend pumpkin seeds and oats until finely ground.
2. Add almond butter, honey, chia seeds, and vanilla extract.
3. Process until mixture forms a sticky dough.
4. Roll into small balls.
5. Refrigerate for at least 30 minutes before serving.

Nutritional Information (per serving):
- Calories: 250
- Carbohydrates: 25g

- Protein: 8g
- Fat: 14g

Detoxifying Green Juice with Dandelion and Parsley

Prep Time: 10 minutes
Portion Size: 2

Ingredients:
- 1 cup chopped dandelion greens
- 1/2 cup chopped parsley
- 1 cucumber, chopped
- 2 celery stalks, chopped
- 1 apple, chopped
- 1/2 lemon, juiced
- 1-inch piece of ginger, peeled and chopped

Instructions:
1. Add all ingredients to a juicer.
2. Juice until smooth.
3. Pour into glasses and serve immediately.

Nutritional Information (per serving):
- Calories: 100
- Carbohydrates: 25g
- Protein: 2g
- Fat: 0g

Spicy Lentil Soup with Turmeric and Ginger for Inflammation Reduction

Prep Time: 15 minutes
Cook Time: 30 minutes
Portion Size: 2

Ingredients:
- 1 tablespoon olive oil
- 1 onion, chopped
- 2 cloves garlic, minced
- 1-inch piece ginger, minced

- 1 cup red lentils, rinsed
- 4 cups vegetable broth
- 1 teaspoon ground turmeric
- 1 teaspoon ground cumin
- 1/2 teaspoon chili powder
- Salt and pepper to taste

Instructions:
1. Heat olive oil in a large pot over medium heat.
2. Sauté onion, garlic, and ginger until softened.
3. Add lentils, broth, turmeric, cumin, chili powder, salt, and pepper.
4. Bring to a boil, then reduce heat and simmer for 30 minutes or until lentils are tender.
5. Serve hot.

Nutritional Information (per serving):
- Calories: 220
- Carbohydrates: 35g
- Protein: 12g
- Fat: 5g

Omega-3 Rich Salmon Salad with Avocado and Walnuts

Prep Time: 15 minutes
Portion Size: 2

Ingredients:
- 2 salmon fillets, cooked and flaked
- 1 avocado, diced
- 1/4 cup walnuts, chopped
- 4 cups mixed greens
- 1/2 cup cherry tomatoes, halved
- 1/4 cup red onion, thinly sliced
- 2 tablespoons olive oil
- 1 tablespoon lemon juice
- Salt and pepper to taste

Instructions:
1. In a large bowl, combine salmon, avocado, walnuts, mixed greens, cherry tomatoes, and red onion.
2. Drizzle with olive oil and lemon juice.
3. Toss gently to combine.
4. Season with salt and pepper to taste.

Nutritional Information (per serving):
- Calories: 400

- Carbohydrates: 12g
- Protein: 28g
- Fat: 30g

Adaptogenic Herbal Tea with Ashwagandha and Holy Basil

Prep Time: 5 minutes
Portion Size: 2

Ingredients:
- 2 cups water
- 1 teaspoon dried ashwagandha root
- 1 teaspoon dried holy basil (tulsi)
- 1 teaspoon dried peppermint
- 1 teaspoon honey

Instructions:
1. Bring water to a boil.
2. Add ashwagandha, holy basil, and peppermint to a teapot or heatproof container.
3. Pour boiling water over the herbs.
4. Steep for 5-10 minutes.
5. Strain the herbs and pour the tea into cups.
6. Add honey to taste.

Nutritional Information (per serving):
- Calories: 10
- Carbohydrates: 2g
- Protein: 0g
- Fat: 0g

Fermented Sauerkraut for Gut Health and Hormone Regulation

Prep Time: 20 minutes
Fermentation Time: 1-2 weeks
Portion Size: 2

Ingredients:
- 1 small head of cabbage, shredded
- 1 tablespoon sea salt
- 1 teaspoon caraway seeds (optional)

Instructions:
1. In a large bowl, combine cabbage and salt.
2. Massage the cabbage for about 10 minutes until it releases its juices.
3. Add caraway seeds if using.
4. Pack the cabbage tightly into a jar, ensuring it is submerged in its liquid.
5. Cover the jar with a cloth and secure with a rubber band.
6. Let ferment at room temperature for 1-2 weeks, checking daily to ensure cabbage remains submerged.
7. Once fermented, refrigerate and consume within a month.

Nutritional Information (per serving):
- Calories: 20
- Carbohydrates: 4g
- Protein: 1g
- Fat: 0g

Sweet Potato and Spinach Frittata for Adrenal Support

Prep Time: 10 minutes
Cook Time: 20 minutes
Portion Size: 2

Ingredients:
- 1 tablespoon olive oil
- 1 small sweet potato, peeled and diced
- 1 cup fresh spinach
- 4 eggs, beaten
- 1/4 cup feta cheese, crumbled
- Salt and pepper to taste

Instructions:
1. Preheat oven to 375°F (190°C).
2. Heat olive oil in an oven-safe skillet over medium heat.
3. Add sweet potato and cook until tender.
4. Add spinach and cook until wilted.
5. Pour beaten eggs over the vegetables in the skillet.
6. Sprinkle with feta cheese, salt, and pepper.
7. Transfer the skillet to the oven and bake for 15 minutes, or until eggs are set.
8. Let cool slightly before serving.

Nutritional Information (per serving):
- Calories: 250
- Carbohydrates: 15g
- Protein: 14g

- Fat: 15g

Hormone-Regulating Golden Milk with Turmeric and Coconut Milk

Prep Time: 5 minutes
Portion Size: 2

Ingredients:
- 2 cups coconut milk
- 1 teaspoon ground turmeric
- 1/2 teaspoon ground cinnamon
- 1/4 teaspoon ground ginger
- 1 tablespoon honey
- 1/2 teaspoon vanilla extract

Instructions:
1. In a saucepan, combine coconut milk, turmeric, cinnamon, and ginger.
2. Heat over medium heat until warm, but do not boil.
3. Stir in honey and vanilla extract.
4. Pour into mugs and serve warm.

Nutritional Information (per serving):
- Calories: 150
- Carbohydrates: 15g
- Protein: 1g
- Fat: 10g

Broccoli and Cauliflower Stir-Fry for Estrogen Detox

Prep Time: 10 minutes
Cook Time: 10 minutes
Portion Size: 2

Ingredients:
- 1 tablespoon coconut oil
- 1 cup broccoli florets
- 1 cup cauliflower florets
- 1 red bell pepper, sliced
- 2 cloves garlic, minced
- 1 tablespoon soy sauce

- 1 teaspoon grated fresh ginger
- 1/4 teaspoon crushed red pepper flakes

Instructions:
1. Heat coconut oil in a large skillet over medium-high heat.
2. Add broccoli, cauliflower, and red bell pepper; stir-fry for 5 minutes.
3. Add garlic, soy sauce, ginger, and red pepper flakes.
4. Continue to stir-fry for another 5 minutes or until vegetables are tender-crisp.
5. Serve hot.

Nutritional Information (per serving):
- Calories: 150
- Carbohydrates: 20g
- Protein: 4g
- Fat: 7g

Bone Broth with Garlic and Rosemary for Immune Support

Prep Time: 10 minutes
Cook Time: 4 hours
Portion Size: 2

Ingredients:
- 4 cups water
- 1 pound beef bones
- 2 cloves garlic, smashed
- 1 sprig fresh rosemary
- 1 tablespoon apple cider vinegar
- Salt and pepper to taste

Instructions:
1. In a large pot, combine water, bones, garlic, rosemary, and apple cider vinegar.
2. Bring to a boil, then reduce heat and simmer for 4 hours.
3. Strain the broth, discarding solids.
4. Season with salt and pepper to taste.
5. Serve warm.

Nutritional Information (per serving):
- Calories: 100
- Carbohydrates: 2g
- Protein: 10g
- Fat: 5g

Quinoa Salad with Beets and Arugula for Liver Detoxification

Prep Time: 15 minutes
Cook Time: 15 minutes
Portion Size: 2

Ingredients:
- 1/2 cup quinoa, rinsed
- 1 cup water
- 1 cup cooked beets, diced
- 2 cups arugula
- 1/4 cup crumbled goat cheese
- 2 tablespoons olive oil
- 1 tablespoon balsamic vinegar
- Salt and pepper to taste

Instructions:
1. In a saucepan, bring quinoa and water to a boil.
2. Reduce heat, cover, and simmer for 15 minutes, or until water is absorbed and quinoa is tender.
3. In a large bowl, combine quinoa, beets, arugula, and goat cheese.
4. Drizzle with olive oil and balsamic vinegar.
5. Toss gently and season with salt and pepper to taste.

Nutritional Information (per serving):
- Calories: 250
- Carbohydrates: 30g
- Protein: 8g
- Fat: 12g

Almond Butter Energy Bars with Seeds and Nuts

Prep Time: 15 minutes
Portion Size: 2

Ingredients:
- 1/2 cup almond butter
- 1/4 cup honey
- 1/2 cup rolled oats
- 1/4 cup chopped almonds
- 1/4 cup pumpkin seeds
- 2 tablespoons chia seeds
- 1 teaspoon vanilla extract

Instructions:
1. In a large bowl, combine almond butter and honey.
2. Stir in oats, almonds, pumpkin seeds, chia seeds, and vanilla extract.
3. Press mixture into a lined baking dish.
4. Refrigerate for at least 1 hour before cutting into bars.

Nutritional Information (per serving):
- Calories: 300
- Carbohydrates: 25g
- Protein: 8g
- Fat: 20g

Coconut and Berry Parfait with Probiotic Yogurt

Prep Time: 10 minutes
Portion Size: 2

Ingredients:
- 1 cup probiotic yogurt
- 1/2 cup fresh berries (strawberries, blueberries, raspberries)
- 1/4 cup unsweetened coconut flakes
- 2 tablespoons chia seeds
- 1 teaspoon honey

Instructions:
1. In two serving glasses, layer yogurt, berries, coconut flakes, and chia seeds.
2. Drizzle with honey.
3. Serve immediately.

Nutritional Information (per serving):
- Calories: 200
- Carbohydrates: 20g
- Protein: 8g
- Fat: 10g

Hormone-Balancing Smoothie Bowl with Spirulina and Banana

Prep Time: 10 minutes
Portion Size: 2

Ingredients:

- 2 bananas, sliced and frozen
- 1 cup almond milk
- 1 tablespoon spirulina powder
- 1/4 cup granola
- 1/4 cup fresh berries
- 1 tablespoon chia seeds

Instructions:

1. In a blender, combine frozen bananas, almond milk, and spirulina powder.
2. Blend until smooth.
3. Pour into bowls and top with granola, berries, and chia seeds.

Nutritional Information (per serving):

- Calories: 250
- Carbohydrates: 45g
- Protein: 6g
- Fat: 6g

Sesame Seed and Honey Bars for Hormone Support

Prep Time: 15 minutes
Portion Size: 2

Ingredients:

- 1/2 cup sesame seeds
- 1/4 cup honey
- 1/4 cup rolled oats
- 1/4 cup chopped almonds
- 1 teaspoon vanilla extract

Instructions:

1. In a large bowl, combine sesame seeds, honey, oats, almonds, and vanilla extract.
2. Press mixture into a lined baking dish.
3. Refrigerate for at least 1 hour before cutting into bars.

Nutritional Information (per serving):

- Calories: 200
- Carbohydrates: 25g
- Protein: 5g
- Fat: 10g

Zucchini Noodles with Pesto for Anti-Inflammatory Benefits

Prep Time: 15 minutes
Portion Size: 2

Ingredients:
- 2 medium zucchinis, spiralized
- 1/4 cup basil pesto
- 1/4 cup cherry tomatoes, halved
- 2 tablespoons pine nuts
- Salt and pepper to taste

Instructions:
1. In a large bowl, toss zucchini noodles with pesto.
2. Add cherry tomatoes and pine nuts.
3. Season with salt and pepper to taste.
4. Serve immediately.

Nutritional Information (per serving):
- Calories: 150
- Carbohydrates: 10g
- Protein: 4g
- Fat: 12g

Herbs for Stress Management and Hormonal Balance

1. **Ashwagandha**: Known for its adaptogenic properties, ashwagandha helps the body cope with stress and supports overall hormonal balance.
2. **Holy Basil (Tulsi)**: This herb helps reduce stress and anxiety while promoting hormonal equilibrium.
3. **Maca Root**: Often used to enhance energy, stamina, and libido, maca root also supports endocrine function.
4. **Rhodiola Rosea**: Known for boosting energy and reducing fatigue, rhodiola rosea helps the body adapt to stress.
5. **Licorice Root**: Supports adrenal function and helps regulate cortisol levels, beneficial for stress management.
6. **Schisandra Berry**: Helps improve concentration and endurance, while also supporting the adrenal glands and balancing hormones.
7. **Ginseng**: Both American and Asian ginseng are known for their ability to enhance vitality, reduce stress, and balance hormones.
8. **Rehmannia**: Often used in traditional Chinese medicine to support adrenal health and hormonal balance.
9. **Vitex (Chaste Tree Berry)**: Supports the pituitary gland and helps balance female hormones, particularly beneficial for PMS and menopause symptoms.
10. **Eleuthero (Siberian Ginseng)**: Enhances the body's ability to handle stress and improves overall energy and stamina.
11. **Lemon Balm**: Known for its calming effects, lemon balm helps reduce stress and anxiety while promoting a sense of well-being.
12. **Passionflower**: Helps reduce anxiety and insomnia, supporting a more balanced hormonal state through improved sleep and relaxation.
13. **Valerian Root**: Known for its sedative properties, valerian root helps with stress management by promoting restful sleep and relaxation.

Natural Recipes for Mental and Brain Health

Brain-Boosting Blueberry and Spinach Smoothie

Prep Time: 5 minutes
Portion Size: 2

Ingredients:
- 1 cup almond milk
- 1 banana
- 1/2 cup fresh blueberries
- 1 cup fresh spinach
- 1 tablespoon chia seeds
- 1 teaspoon honey
- 1/2 teaspoon vanilla extract

Instructions:
1. Combine all ingredients in a blender.
2. Blend until smooth.
3. Pour into glasses and serve immediately.

Nutritional Information (per serving):
- Calories: 150
- Carbohydrates: 30g
- Protein: 3g
- Fat: 3g

Walnut and Berry Salad with Mixed Greens

Prep Time: 10 minutes
Portion Size: 2

Ingredients:
- 4 cups mixed greens
- 1/2 cup fresh berries (strawberries, blueberries, raspberries)
- 1/4 cup walnuts, chopped
- 2 tablespoons feta cheese, crumbled
- 2 tablespoons balsamic vinaigrette

Instructions:
1. In a large bowl, combine mixed greens, berries, walnuts, and feta cheese.
2. Drizzle with balsamic vinaigrette.
3. Toss gently and serve.

Nutritional Information (per serving):
- Calories: 200
- Carbohydrates: 15g
- Protein: 5g
- Fat: 15g

Turmeric and Black Pepper Golden Milk for Cognitive Health

Prep Time: 5 minutes
Portion Size: 2

Ingredients:
- 2 cups coconut milk
- 1 teaspoon ground turmeric
- 1/4 teaspoon black pepper
- 1/2 teaspoon ground cinnamon
- 1 tablespoon honey

Instructions:
1. In a saucepan, combine coconut milk, turmeric, black pepper, and cinnamon.
2. Heat over medium heat until warm, but do not boil.
3. Stir in honey.
4. Pour into mugs and serve warm.

Nutritional Information (per serving):
- Calories: 150
- Carbohydrates: 18g
- Protein: 1g
- Fat: 10g

Omega-3 Rich Chia Seed Pudding with Almond Milk

Prep Time: 10 minutes (plus overnight refrigeration)
Portion Size: 2

Ingredients:
- 1 cup almond milk
- 3 tablespoons chia seeds
- 1 teaspoon vanilla extract
- 1 tablespoon honey

- 1/2 cup fresh berries

Instructions:
1. In a bowl, combine almond milk, chia seeds, vanilla extract, and honey.
2. Stir well to combine.
3. Cover and refrigerate overnight.
4. Stir again before serving, and top with fresh berries.

Nutritional Information (per serving):
- Calories: 180
- Carbohydrates: 25g
- Protein: 4g
- Fat: 9g

Dark Chocolate and Almond Energy Bars for Mental Clarity

Prep Time: 15 minutes
Portion Size: 2

Ingredients:
- 1/2 cup rolled oats
- 1/4 cup almond butter
- 1/4 cup honey
- 1/4 cup dark chocolate chips
- 1/4 cup chopped almonds
- 1 tablespoon chia seeds
- 1 teaspoon vanilla extract

Instructions:
1. In a large bowl, combine rolled oats, almond butter, honey, dark chocolate chips, almonds, chia seeds, and vanilla extract.
2. Press mixture into a lined baking dish.
3. Refrigerate for at least 1 hour before cutting into bars.

Nutritional Information (per serving):
- Calories: 250
- Carbohydrates: 30g
- Protein: 6g
- Fat: 12g

Roasted Beet and Goat Cheese Salad for Enhanced Memory

Prep Time: 10 minutes
Cook Time: 30 minutes
Portion Size: 2

Ingredients:
- 2 medium beets, roasted and diced
- 4 cups mixed greens
- 1/4 cup goat cheese, crumbled
- 2 tablespoons walnuts, chopped
- 2 tablespoons balsamic vinaigrette

Instructions:
1. In a large bowl, combine roasted beets, mixed greens, goat cheese, and walnuts.
2. Drizzle with balsamic vinaigrette.
3. Toss gently and serve.

Nutritional Information (per serving):
- Calories: 220
- Carbohydrates: 18g
- Protein: 6g
- Fat: 14g

Avocado and Tomato Brain-Healthy Toast

Prep Time: 10 minutes
Portion Size: 2

Ingredients:
- 2 slices whole-grain bread, toasted
- 1 ripe avocado
- 1/2 cup cherry tomatoes, halved
- 1 tablespoon olive oil
- Salt and pepper to taste
- 1 teaspoon lemon juice

Instructions:
1. Mash the avocado in a bowl and add lemon juice, salt, and pepper.
2. Spread the avocado mixture on the toasted bread.
3. Top with cherry tomatoes.
4. Drizzle with olive oil and serve.

Nutritional Information (per serving):
- Calories: 250
- Carbohydrates: 28g
- Protein: 5g
- Fat: 15g

Ginkgo Biloba Herbal Tea for Focus and Concentration

Prep Time: 5 minutes
Portion Size: 2

Ingredients:
- 2 cups water
- 1 tablespoon dried ginkgo biloba leaves
- 1 teaspoon honey (optional)

Instructions:
1. Bring water to a boil.
2. Add dried ginkgo biloba leaves to a teapot or heatproof container.
3. Pour boiling water over the leaves.
4. Steep for 5-10 minutes.
5. Strain the leaves and pour the tea into cups.
6. Add honey if desired.

Nutritional Information (per serving):
- Calories: 10
- Carbohydrates: 2g
- Protein: 0g
- Fat: 0g

Salmon and Quinoa Power Bowl with Leafy Greens

Prep Time: 15 minutes
Cook Time: 20 minutes
Portion Size: 2

Ingredients:
- 2 salmon fillets
- 1/2 cup quinoa, rinsed
- 1 cup water
- 2 cups mixed greens
- 1/2 avocado, sliced
- 1/4 cup cherry tomatoes, halved
- 2 tablespoons olive oil
- 1 tablespoon lemon juice

- Salt and pepper to taste

Instructions:
1. Preheat oven to 375°F (190°C).
2. Season salmon fillets with salt and pepper and bake for 20 minutes or until cooked through.
3. In a saucepan, bring quinoa and water to a boil. Reduce heat, cover, and simmer for 15 minutes or until water is absorbed.
4. In a large bowl, combine cooked quinoa, mixed greens, avocado, and cherry tomatoes.
5. Drizzle with olive oil and lemon juice. Toss gently.
6. Top with the baked salmon and serve.

Nutritional Information (per serving):
- Calories: 400
- Carbohydrates: 25g
- Protein: 28g
- Fat: 20g

Turmeric and Ginger Spiced Carrot Soup

Prep Time: 15 minutes
Cook Time: 30 minutes
Portion Size: 2

Ingredients:
- 1 tablespoon olive oil
- 1 onion, chopped
- 2 cloves garlic, minced
- 1-inch piece ginger, minced
- 4 carrots, peeled and chopped
- 3 cups vegetable broth
- 1 teaspoon ground turmeric
- Salt and pepper to taste

Instructions:
1. Heat olive oil in a large pot over medium heat.
2. Add onion, garlic, and ginger; sauté until softened.
3. Add carrots, vegetable broth, and turmeric.
4. Bring to a boil, then reduce heat and simmer for 20 minutes or until carrots are tender.
5. Blend soup until smooth using an immersion blender.
6. Season with salt and pepper to taste and serve warm.

Nutritional Information (per serving):
- Calories: 150
- Carbohydrates: 28g

- Protein: 2g
- Fat: 6g

Green Tea and Lemon Detox Drink for Mental Alertness

Prep Time: 5 minutes
Portion Size: 2

Ingredients:
- 2 cups water
- 2 green tea bags
- 1/2 lemon, juiced
- 1 teaspoon honey (optional)

Instructions:
1. Boil water and steep green tea bags for 3-5 minutes.
2. Remove tea bags and add lemon juice.
3. Stir in honey if desired.
4. Serve warm or chilled.

Nutritional Information (per serving):
- Calories: 5
- Carbohydrates: 1g
- Protein: 0g
- Fat: 0g

Flaxseed and Berry Yogurt Parfait for Brain Health

Prep Time: 10 minutes
Portion Size: 2

Ingredients:
- 1 cup plain Greek yogurt
- 2 tablespoons ground flaxseed
- 1/2 cup fresh berries (blueberries, strawberries, raspberries)
- 1 tablespoon honey

Instructions:
1. In two serving glasses, layer yogurt, ground flaxseed, and fresh berries.
2. Drizzle with honey.

3. Serve immediately.

Nutritional Information (per serving):
- Calories: 200
- Carbohydrates: 20g
- Protein: 14g
- Fat: 8g

Rosemary and Olive Oil Roasted Sweet Potatoes

Prep Time: 10 minutes
Cook Time: 30 minutes
Portion Size: 2

Ingredients:
- 2 medium sweet potatoes, peeled and diced
- 1 tablespoon olive oil
- 1 teaspoon dried rosemary
- Salt and pepper to taste

Instructions:
1. Preheat oven to 400°F (200°C).
2. In a bowl, toss sweet potatoes with olive oil, rosemary, salt, and pepper.
3. Spread sweet potatoes on a baking sheet in a single layer.
4. Roast for 30 minutes or until tender and slightly browned.
5. Serve warm.

Nutritional Information (per serving):
- Calories: 180
- Carbohydrates: 32g
- Protein: 2g
- Fat: 6g

Coconut and Berry Smoothie Bowl for Cognitive Function

Prep Time: 10 minutes
Portion Size: 2

Ingredients:
- 1 banana, sliced and frozen

- 1 cup coconut milk
- 1/2 cup fresh berries (blueberries, strawberries, raspberries)
- 1/4 cup granola
- 1 tablespoon chia seeds
- 1 tablespoon shredded coconut

Instructions:

1. In a blender, combine frozen banana and coconut milk.
2. Blend until smooth.
3. Pour into bowls and top with fresh berries, granola, chia seeds, and shredded coconut.
4. Serve immediately.

Nutritional Information (per serving):
- Calories: 250
- Carbohydrates: 35g
- Protein: 4g
- Fat: 12g

Pumpkin Seed and Spinach Pesto Pasta

Prep Time: 15 minutes
Cook Time: 10 minutes
Portion Size: 2

Ingredients:
- 4 ounces whole wheat pasta
- 1 cup fresh spinach
- 1/4 cup pumpkin seeds
- 1/4 cup grated Parmesan cheese
- 1 clove garlic
- 1/4 cup olive oil
- Salt and pepper to taste

Instructions:

1. Cook pasta according to package instructions. Drain and set aside.
2. In a food processor, combine spinach, pumpkin seeds, Parmesan cheese, and garlic.
3. Pulse until finely chopped.
4. With the processor running, gradually add olive oil until the mixture is smooth.
5. Toss pasta with pesto sauce.
6. Season with salt and pepper to taste.
7. Serve warm.

Nutritional Information (per serving):
- Calories: 400

- Carbohydrates: 45g
- Protein: 10g
- Fat: 20g

Memory-Enhancing Ginseng and Honey Herbal Infusion

Prep Time: 5 minutes
Portion Size: 2

Ingredients:
- 2 cups water
- 1 tablespoon dried ginseng root
- 1 teaspoon honey

Instructions:
1. Bring water to a boil.
2. Add dried ginseng root to a teapot or heatproof container.
3. Pour boiling water over the ginseng root.
4. Steep for 5-10 minutes.
5. Strain the ginseng root and pour the tea into cups.
6. Add honey and stir.

Nutritional Information (per serving):
- Calories: 15
- Carbohydrates: 4g
- Protein: 0g
- Fat: 0g

Dark Leafy Greens and Citrus Salad with Walnuts

Prep Time: 10 minutes
Portion Size: 2

Ingredients:
- 4 cups mixed dark leafy greens (kale, spinach, arugula)
- 1 orange, peeled and segmented
- 1/4 cup walnuts, chopped
- 2 tablespoons olive oil
- 1 tablespoon lemon juice
- Salt and pepper to taste

Instructions:
1. In a large bowl, combine dark leafy greens, orange segments, and walnuts.
2. Drizzle with olive oil and lemon juice.
3. Toss gently to combine.
4. Season with salt and pepper to taste.
5. Serve immediately.

Nutritional Information (per serving):
- Calories: 200
- Carbohydrates: 18g
- Protein: 4g
- Fat: 14g

Baked Salmon with Garlic and Rosemary

Prep Time: 10 minutes
Cook Time: 20 minutes
Portion Size: 2

Ingredients:
- 2 salmon fillets
- 2 cloves garlic, minced
- 1 tablespoon olive oil
- 1 teaspoon dried rosemary
- Salt and pepper to taste

Instructions:
1. Preheat oven to 375°F (190°C).
2. Place salmon fillets on a baking sheet.
3. In a small bowl, combine garlic, olive oil, rosemary, salt, and pepper.
4. Spread the garlic mixture over the salmon fillets.
5. Bake for 20 minutes or until salmon is cooked through.
6. Serve warm.

Nutritional Information (per serving):
- Calories: 350
- Carbohydrates: 2g
- Protein: 34g
- Fat: 22g

Nutty Quinoa and Spinach Salad for Mental Clarity

Prep Time: 15 minutes
Cook Time: 15 minutes
Portion Size: 2

Ingredients:
- 1/2 cup quinoa, rinsed
- 1 cup water
- 2 cups fresh spinach
- 1/4 cup chopped nuts (almonds, walnuts, pecans)
- 1/4 cup crumbled feta cheese
- 2 tablespoons olive oil
- 1 tablespoon balsamic vinegar
- Salt and pepper to taste

Instructions:
1. In a saucepan, bring quinoa and water to a boil. Reduce heat, cover, and simmer for 15 minutes or until water is absorbed.
2. In a large bowl, combine cooked quinoa, spinach, nuts, and feta cheese.
3. Drizzle with olive oil and balsamic vinegar.
4. Toss gently to combine.
5. Season with salt and pepper to taste.
6. Serve warm or chilled.

Nutritional Information (per serving):
- Calories: 300
- Carbohydrates: 25g
- Protein: 10g
- Fat: 18g

Cacao and Almond Brain-Boosting Smoothie

Prep Time: 5 minutes
Portion Size: 2

Ingredients:
- 1 banana
- 1 cup almond milk
- 1 tablespoon raw cacao powder
- 1 tablespoon almond butter
- 1 teaspoon honey

- 1/2 teaspoon vanilla extract

Instructions:

1. Combine all ingredients in a blender.
2. Blend until smooth.
3. Pour into glasses and serve immediately.

Nutritional Information (per serving):

- Calories: 180
- Carbohydrates: 28g
- Protein: 4g
- Fat: 8g

Nootropic and Adaptogenic Herbs for Brain Health

1. **Ginkgo Biloba**: Known to enhance cognitive function, improve memory, and boost mental clarity by increasing blood flow to the brain.
2. **Rhodiola Rosea**: An adaptogen that helps reduce fatigue, improve mood, and enhance mental performance under stress.
3. **Ashwagandha**: Supports brain health by reducing anxiety, improving focus, and enhancing overall cognitive function.
4. **Bacopa Monnieri**: Traditionally used in Ayurvedic medicine to improve memory, learning ability, and concentration.
5. **Lion's Mane Mushroom**: Promotes nerve growth factor production, which supports brain health and cognitive function.
6. **Panax Ginseng**: Boosts energy, enhances mental performance, and helps the body adapt to stress.
7. **Gotu Kola**: Known to improve memory, cognitive function, and reduce anxiety.
8. **Holy Basil (Tulsi)**: Reduces stress, enhances mental clarity, and supports overall brain function.
9. **Maca Root**: Increases energy, stamina, and mental focus while balancing hormones.
10. **Cordyceps**: Improves mental energy and focus, supports overall brain health.
11. **L-Theanine**: Found in green tea, promotes relaxation without drowsiness, and enhances cognitive performance.
12. **Eleuthero (Siberian Ginseng)**: Enhances mental clarity, reduces stress, and supports overall brain health.
13. **Schisandra Berry**: Improves concentration, endurance, and helps the body cope with stress.
14. **Rosemary**: Contains compounds that improve memory and concentration by increasing blood flow to the brain.
15. **Sage**: Known for its ability to enhance memory and cognitive function.

Natural Recipes for Physical Recovery

Anti-Inflammatory Turmeric and Ginger Smoothie

Prep Time: 5 minutes
Portion Size: 2

Ingredients:
- 1 cup coconut milk
- 1 banana
- 1/2 teaspoon ground turmeric
- 1/2 teaspoon ground ginger
- 1 tablespoon honey
- 1/2 cup ice

Instructions:
1. Combine all ingredients in a blender.
2. Blend until smooth.
3. Pour into glasses and serve immediately.

Nutritional Information (per serving):
- Calories: 120
- Carbohydrates: 25g
- Protein: 1g
- Fat: 4g

Cherry and Almond Recovery Smoothie

Prep Time: 5 minutes
Portion Size: 2

Ingredients:
- 1 cup almond milk
- 1 cup frozen cherries
- 1 banana
- 1 tablespoon almond butter
- 1 teaspoon honey

Instructions:
1. Combine all ingredients in a blender.
2. Blend until smooth.
3. Pour into glasses and serve immediately.

Nutritional Information (per serving):

- Calories: 180
- Carbohydrates: 35g
- Protein: 3g
- Fat: 6g

Protein-Packed Quinoa and Black Bean Salad

Prep Time: 15 minutes
Cook Time: 15 minutes
Portion Size: 2

Ingredients:
- 1/2 cup quinoa, rinsed
- 1 cup water
- 1 cup black beans, rinsed and drained
- 1/2 cup corn kernels
- 1/2 red bell pepper, diced
- 1/4 cup red onion, diced
- 1/4 cup chopped cilantro
- 2 tablespoons olive oil
- 1 tablespoon lime juice
- Salt and pepper to taste

Instructions:
1. In a saucepan, bring quinoa and water to a boil. Reduce heat, cover, and simmer for 15 minutes or until water is absorbed.
2. In a large bowl, combine cooked quinoa, black beans, corn, bell pepper, red onion, and cilantro.
3. Drizzle with olive oil and lime juice.
4. Toss gently to combine and season with salt and pepper to taste.

Nutritional Information (per serving):
- Calories: 300
- Carbohydrates: 45g
- Protein: 10g
- Fat: 10g

Omega-3 Rich Salmon with Avocado Salsa

Prep Time: 10 minutes
Cook Time: 15 minutes

Portion Size: 2

Ingredients:
- 2 salmon fillets
- 1 avocado, diced
- 1/2 cup cherry tomatoes, halved
- 1/4 red onion, finely chopped
- 1 tablespoon lime juice
- 1 tablespoon olive oil
- Salt and pepper to taste

Instructions:
1. Preheat oven to 375°F (190°C).
2. Season salmon fillets with salt and pepper and place on a baking sheet.
3. Bake for 15 minutes or until cooked through.
4. In a bowl, combine avocado, cherry tomatoes, red onion, lime juice, and olive oil.
5. Serve the salmon topped with avocado salsa.

Nutritional Information (per serving):
- Calories: 400
- Carbohydrates: 10g
- Protein: 28g
- Fat: 28g

Baked Sweet Potato and Spinach Frittata

Prep Time: 10 minutes
Cook Time: 20 minutes
Portion Size: 2

Ingredients:
- 1 tablespoon olive oil
- 1 small sweet potato, peeled and diced
- 1 cup fresh spinach
- 4 eggs, beaten
- 1/4 cup feta cheese, crumbled
- Salt and pepper to taste

Instructions:
1. Preheat oven to 375°F (190°C).
2. Heat olive oil in an oven-safe skillet over medium heat.
3. Add sweet potato and cook until tender.
4. Add spinach and cook until wilted.
5. Pour beaten eggs over the vegetables in the skillet.

6. Sprinkle with feta cheese, salt, and pepper.
7. Transfer the skillet to the oven and bake for 15 minutes, or until eggs are set.
8. Let cool slightly before serving.

Nutritional Information (per serving):
- Calories: 250
- Carbohydrates: 15g
- Protein: 14g
- Fat: 15g

Rehydrating Coconut Water and Pineapple Drink

Prep Time: 5 minutes
Portion Size: 2

Ingredients:
- 2 cups coconut water
- 1 cup pineapple chunks
- 1 tablespoon lime juice
- 1/2 cup ice

Instructions:
1. Combine all ingredients in a blender.
2. Blend until smooth.
3. Pour into glasses and serve immediately.

Nutritional Information (per serving):
- Calories: 80
- Carbohydrates: 20g
- Protein: 1g
- Fat: 0g

Muscle-Repairing Chicken and Broccoli Stir-Fry

Prep Time: 10 minutes
Cook Time: 15 minutes
Portion Size: 2

Ingredients:
- 1 tablespoon olive oil
- 2 chicken breasts, sliced

- 2 cups broccoli florets
- 1 red bell pepper, sliced
- 2 cloves garlic, minced
- 1 tablespoon soy sauce
- 1 teaspoon grated ginger
- Salt and pepper to taste

Instructions:

1. Heat olive oil in a large skillet over medium-high heat.
2. Add chicken and cook until browned.
3. Add broccoli, bell pepper, garlic, soy sauce, and ginger.
4. Stir-fry for 5-7 minutes or until vegetables are tender-crisp.
5. Season with salt and pepper to taste.

Nutritional Information (per serving):

- Calories: 300
- Carbohydrates: 10g
- Protein: 35g
- Fat: 12g

Beetroot and Carrot Juice for Endurance

Prep Time: 10 minutes
Portion Size: 2

Ingredients:

- 2 medium beets, peeled and chopped
- 4 carrots, peeled and chopped
- 1 apple, cored and chopped
- 1-inch piece of ginger, peeled

Instructions:

1. Add all ingredients to a juicer.
2. Juice until smooth.
3. Pour into glasses and serve immediately.

Nutritional Information (per serving):

- Calories: 100
- Carbohydrates: 25g
- Protein: 2g
- Fat: 0g

Protein-Packed Greek Yogurt with Berries and Nuts

Prep Time: 5 minutes
Portion Size: 2

Ingredients:
- 1 cup plain Greek yogurt
- 1/2 cup mixed berries (blueberries, strawberries, raspberries)
- 1/4 cup chopped nuts (almonds, walnuts, pecans)
- 1 tablespoon honey

Instructions:
1. Divide Greek yogurt between two bowls.
2. Top with mixed berries and chopped nuts.
3. Drizzle with honey.
4. Serve immediately.

Nutritional Information (per serving):
- Calories: 200
- Carbohydrates: 25g
- Protein: 12g
- Fat: 8g

Anti-Fatigue Green Tea and Lemon Cooler

Prep Time: 5 minutes
Portion Size: 2

Ingredients:
- 2 cups water
- 2 green tea bags
- 1/2 lemon, juiced
- 1 teaspoon honey
- Ice cubes

Instructions:
1. Boil water and steep green tea bags for 3-5 minutes.
2. Remove tea bags and let the tea cool.
3. Add lemon juice and honey.
4. Serve over ice.

Nutritional Information (per serving):
- Calories: 10

- Carbohydrates: 2g
- Protein: 0g
- Fat: 0g

Pumpkin Seed and Oat Energy Bites

Prep Time: 15 minutes
Portion Size: 2

Ingredients:
- 1/2 cup rolled oats
- 1/4 cup pumpkin seeds
- 1/4 cup almond butter
- 1/4 cup honey
- 1 tablespoon chia seeds
- 1 teaspoon vanilla extract

Instructions:
1. In a large bowl, combine rolled oats, pumpkin seeds, almond butter, honey, chia seeds, and vanilla extract.
2. Roll the mixture into small balls.
3. Refrigerate for at least 30 minutes before serving.

Nutritional Information (per serving):
- Calories: 250
- Carbohydrates: 30g
- Protein: 8g
- Fat: 12g

Banana and Peanut Butter Recovery Smoothie

Prep Time: 5 minutes
Portion Size: 2

Ingredients:
- 2 bananas
- 1 cup almond milk
- 2 tablespoons peanut butter
- 1 tablespoon honey
- 1/2 cup ice

Instructions:
1. Combine all ingredients in a blender.
2. Blend until smooth.
3. Pour into glasses and serve immediately.

Nutritional Information (per serving):
- Calories: 220
- Carbohydrates: 30g
- Protein: 6g
- Fat: 9g

Hearty Lentil and Vegetable Stew

Prep Time: 15 minutes
Cook Time: 30 minutes
Portion Size: 2

Ingredients:
- 1 tablespoon olive oil
- 1 onion, chopped
- 2 cloves garlic, minced
- 1 carrot, diced
- 1 celery stalk, diced
- 1 cup lentils, rinsed
- 4 cups vegetable broth
- 1 can diced tomatoes
- 1 teaspoon dried thyme
- 1 bay leaf
- Salt and pepper to taste

Instructions:
1. Heat olive oil in a large pot over medium heat.
2. Add onion, garlic, carrot, and celery; sauté until softened.
3. Add lentils, vegetable broth, diced tomatoes, thyme, and bay leaf.
4. Bring to a boil, then reduce heat and simmer for 30 minutes or until lentils are tender.
5. Season with salt and pepper to taste.
6. Remove bay leaf before serving.

Nutritional Information (per serving):
- Calories: 250
- Carbohydrates: 40g
- Protein: 12g
- Fat: 6g

Chia Seed Pudding with Coconut Milk and Mango

Prep Time: 10 minutes (plus overnight refrigeration)
Portion Size: 2

Ingredients:
- 1 cup coconut milk
- 3 tablespoons chia seeds
- 1 teaspoon vanilla extract
- 1 tablespoon honey
- 1/2 cup diced mango

Instructions:
1. In a bowl, combine coconut milk, chia seeds, vanilla extract, and honey.
2. Stir well to combine.
3. Cover and refrigerate overnight.
4. Stir again before serving, and top with diced mango.

Nutritional Information (per serving):
- Calories: 200
- Carbohydrates: 25g
- Protein: 4g
- Fat: 10g

Tropical Mango and Chia Recovery Drink

Prep Time: 5 minutes
Portion Size: 2

Ingredients:
- 1 cup coconut water
- 1 cup mango chunks
- 1 tablespoon chia seeds
- 1 teaspoon lime juice
- 1/2 cup ice

Instructions:
1. Combine all ingredients in a blender.
2. Blend until smooth.
3. Pour into glasses and serve immediately.

Nutritional Information (per serving):
- Calories: 100

- Carbohydrates: 25g
- Protein: 1g
- Fat: 1g

Walnut and Spinach Pesto Pasta

Prep Time: 15 minutes
Cook Time: 10 minutes
Portion Size: 2

Ingredients:
- 4 ounces whole wheat pasta
- 1 cup fresh spinach
- 1/4 cup walnuts
- 1/4 cup grated Parmesan cheese
- 1 clove garlic
- 1/4 cup olive oil
- Salt and pepper to taste

Instructions:
1. Cook pasta according to package instructions. Drain and set aside.
2. In a food processor, combine spinach, walnuts, Parmesan cheese, and garlic.
3. Pulse until finely chopped.
4. With the processor running, gradually add olive oil until the mixture is smooth.
5. Toss pasta with pesto sauce.
6. Season with salt and pepper to taste.
7. Serve warm.

Nutritional Information (per serving):
- Calories: 350
- Carbohydrates: 40g
- Protein: 10g
- Fat: 18g

Hydrating Watermelon and Mint Salad

Prep Time: 10 minutes
Portion Size: 2

Ingredients:

- 2 cups watermelon, diced
- 1/4 cup fresh mint leaves, chopped
- 1 tablespoon lime juice
- 1 teaspoon honey

Instructions:

1. In a large bowl, combine watermelon and mint leaves.
2. Drizzle with lime juice and honey.
3. Toss gently to combine.
4. Serve immediately.

Nutritional Information (per serving):
- Calories: 80
- Carbohydrates: 21g
- Protein: 1g
- Fat: 0g

Anti-Inflammatory Golden Milk Latte

Prep Time: 5 minutes
Portion Size: 2

Ingredients:
- 2 cups almond milk
- 1 teaspoon ground turmeric
- 1/4 teaspoon black pepper
- 1/2 teaspoon ground cinnamon
- 1 tablespoon honey

Instructions:

1. In a saucepan, combine almond milk, turmeric, black pepper, and cinnamon.
2. Heat over medium heat until warm, but do not boil.
3. Stir in honey.
4. Pour into mugs and serve warm.

Nutritional Information (per serving):
- Calories: 100
- Carbohydrates: 15g
- Protein: 2g
- Fat: 3g

Blueberry and Kale Smoothie Bowl

Prep Time: 10 minutes
Portion Size: 2

Ingredients:
- 1 banana, sliced and frozen
- 1 cup almond milk
- 1/2 cup fresh blueberries
- 1 cup fresh kale
- 1/4 cup granola
- 1 tablespoon chia seeds

Instructions:
1. In a blender, combine frozen banana, almond milk, blueberries, and kale.
2. Blend until smooth.
3. Pour into bowls and top with granola and chia seeds.
4. Serve immediately.

Nutritional Information (per serving):
- Calories: 200
- Carbohydrates: 38g
- Protein: 5g
- Fat: 4g

Spicy Ginger and Turmeric Carrot Soup

Prep Time: 10 minutes
Cook Time: 25 minutes
Portion Size: 2

Ingredients:
- 1 tablespoon olive oil
- 1 onion, chopped
- 2 cloves garlic, minced
- 1-inch piece ginger, minced
- 4 carrots, peeled and chopped
- 3 cups vegetable broth
- 1 teaspoon ground turmeric
- Salt and pepper to taste

Instructions:
1. Heat olive oil in a large pot over medium heat.

2. Add onion, garlic, and ginger; sauté until softened.
3. Add carrots, vegetable broth, and turmeric.
4. Bring to a boil, then reduce heat and simmer for 20 minutes or until carrots are tender.
5. Blend soup until smooth using an immersion blender.
6. Season with salt and pepper to taste and serve warm.

Nutritional Information (per serving):
- Calories: 150
- Carbohydrates: 28g
- Protein: 2g
- Fat: 6g

Natural Supplements and Foods for Athletes

1. **Whey Protein**: A high-quality protein source that supports muscle repair and growth.
2. **Creatine**: Enhances strength, power, and muscle mass.
3. **BCAAs (Branched-Chain Amino Acids)**: Helps reduce muscle soreness and fatigue.
4. **Omega-3 Fatty Acids**: Reduces inflammation and supports cardiovascular health.
5. **Magnesium**: Essential for muscle function, energy production, and reducing cramps.
6. **Turmeric**: Contains curcumin, which has powerful anti-inflammatory properties.
7. **Beetroot Juice**: Improves endurance and reduces muscle fatigue.
8. **Chia Seeds**: Provides sustained energy and hydration.
9. **Coconut Water**: Natural electrolyte drink for hydration.
10. **Tart Cherry Juice**: Reduces muscle soreness and speeds up recovery.
11. **Spirulina**: A nutrient-dense algae that boosts endurance and muscle recovery.
12. **Collagen**: Supports joint health and reduces the risk of injury.
13. **Ginseng**: Enhances physical performance and reduces fatigue.
14. **Ashwagandha**: Increases strength, muscle mass, and reduces stress.
15. **Maca Root**: Boosts energy, stamina, and overall athletic performance.
16. **Green Tea Extract**: Provides antioxidants and boosts metabolism.
17. **Quinoa**: A complete protein that provides sustained energy and muscle repair.
18. **Pumpkin Seeds**: Rich in magnesium and zinc, supporting muscle function and recovery.
19. **Sweet Potatoes**: High in complex carbohydrates and beta-carotene for energy and recovery.
20. **Greek Yogurt**: High in protein and probiotics for muscle repair and gut health.

Natural Recipes for Gut Health

Fermented Sauerkraut for Gut Health

Prep Time: 20 minutes
Fermentation Time: 1-2 weeks
Portion Size: 2

Ingredients:
- 1 small head of cabbage, shredded
- 1 tablespoon sea salt
- 1 teaspoon caraway seeds (optional)

Instructions:
1. In a large bowl, combine cabbage and salt.
2. Massage the cabbage for about 10 minutes until it releases its juices.
3. Add caraway seeds if using.
4. Pack the cabbage tightly into a jar, ensuring it is submerged in its liquid.
5. Cover the jar with a cloth and secure with a rubber band.
6. Let ferment at room temperature for 1-2 weeks, checking daily to ensure cabbage remains submerged.
7. Once fermented, refrigerate and consume within a month.

Nutritional Information (per serving):
- Calories: 20
- Carbohydrates: 4g
- Protein: 1g
- Fat: 0g

Probiotic-Rich Kimchi

Prep Time: 30 minutes
Fermentation Time: 1-2 weeks
Portion Size: 2

Ingredients:
- 1 small napa cabbage, chopped
- 1 tablespoon sea salt
- 2 carrots, julienned
- 2 green onions, chopped
- 1 tablespoon grated ginger
- 2 cloves garlic, minced
- 1 tablespoon fish sauce
- 1 tablespoon Korean red pepper flakes

Instructions:
1. In a large bowl, combine cabbage and salt. Let sit for 2 hours.
2. Rinse cabbage and drain.
3. Add carrots, green onions, ginger, garlic, fish sauce, and red pepper flakes.
4. Mix thoroughly and pack tightly into a jar.
5. Press down to ensure the vegetables are submerged in their liquid.
6. Cover with a cloth and let ferment at room temperature for 1-2 weeks.
7. Refrigerate and consume within a month.

Nutritional Information (per serving):
- Calories: 30
- Carbohydrates: 6g
- Protein: 2g
- Fat: 0g

Kefir Smoothie with Berries and Honey

Prep Time: 5 minutes
Portion Size: 2

Ingredients:
- 1 cup kefir
- 1 cup mixed berries (blueberries, strawberries, raspberries)
- 1 banana
- 1 tablespoon honey
- 1/2 cup ice

Instructions:
1. Combine all ingredients in a blender.
2. Blend until smooth.
3. Pour into glasses and serve immediately.

Nutritional Information (per serving):
- Calories: 150
- Carbohydrates: 30g
- Protein: 4g
- Fat: 2g

Homemade Kombucha Tea

Prep Time: 15 minutes
Fermentation Time: 1-2 weeks
Portion Size: 2

Ingredients:
- 4 cups water
- 2 black tea bags
- 1/4 cup sugar
- 1/2 cup starter kombucha
- 1 SCOBY

Instructions:
1. Boil water and steep tea bags for 10 minutes.
2. Remove tea bags and stir in sugar until dissolved. Let cool to room temperature.
3. Pour tea into a jar, add starter kombucha and SCOBY.
4. Cover with a cloth and secure with a rubber band.
5. Let ferment at room temperature for 1-2 weeks.
6. Remove SCOBY and reserve for next batch.
7. Bottle kombucha and refrigerate.

Nutritional Information (per serving):
- Calories: 30
- Carbohydrates: 7g
- Protein: 0g
- Fat: 0g

Bone Broth with Ginger and Turmeric

Prep Time: 10 minutes
Cook Time: 4 hours
Portion Size: 2

Ingredients:
- 4 cups water
- 1 pound beef bones
- 2 cloves garlic, smashed
- 1-inch piece ginger, sliced
- 1 teaspoon ground turmeric
- 1 tablespoon apple cider vinegar
- Salt and pepper to taste

Instructions:
1. In a large pot, combine water, bones, garlic, ginger, turmeric, and apple cider vinegar.
2. Bring to a boil, then reduce heat and simmer for 4 hours.
3. Strain the broth, discarding solids.
4. Season with salt and pepper to taste.
5. Serve warm.

Nutritional Information (per serving):
- Calories: 100
- Carbohydrates: 2g
- Protein: 10g
- Fat: 5g

Greek Yogurt and Flaxseed Parfait

Prep Time: 5 minutes
Portion Size: 2

Ingredients:
- 1 cup plain Greek yogurt
- 2 tablespoons ground flaxseed
- 1/2 cup fresh berries
- 1 tablespoon honey

Instructions:
1. In two serving glasses, layer Greek yogurt, ground flaxseed, and fresh berries.
2. Drizzle with honey.
3. Serve immediately.

Nutritional Information (per serving):
- Calories: 200
- Carbohydrates: 20g
- Protein: 14g
- Fat: 8g

Miso Soup with Tofu and Seaweed

Prep Time: 10 minutes
Cook Time: 10 minutes
Portion Size: 2

Ingredients:
- 4 cups water
- 2 tablespoons miso paste
- 1/2 cup tofu, cubed
- 1/4 cup seaweed, soaked and drained
- 1 green onion, sliced

Instructions:
1. In a pot, bring water to a boil.
2. Reduce heat to low and dissolve miso paste in the water.
3. Add tofu and seaweed; simmer for 5 minutes.
4. Garnish with sliced green onion and serve.

Nutritional Information (per serving):
- Calories: 70
- Carbohydrates: 6g
- Protein: 6g
- Fat: 2g

Garlic and Herb Marinated Olives

Prep Time: 10 minutes
Marination Time: 1 hour
Portion Size: 2

Ingredients:
- 1 cup mixed olives
- 2 cloves garlic, minced
- 1 tablespoon olive oil
- 1 teaspoon dried oregano
- 1 teaspoon dried thyme
- 1/2 teaspoon red pepper flakes

Instructions:
1. In a bowl, combine olives, garlic, olive oil, oregano, thyme, and red pepper flakes.
2. Mix well and let marinate for at least 1 hour.
3. Serve as a snack or appetizer.

Nutritional Information (per serving):
- Calories: 100
- Carbohydrates: 2g
- Protein: 1g
- Fat: 10g

Roasted Asparagus with Lemon and Garlic

Prep Time: 5 minutes
Cook Time: 15 minutes
Portion Size: 2

Ingredients:
- 1 bunch asparagus, trimmed
- 1 tablespoon olive oil
- 2 cloves garlic, minced
- 1 tablespoon lemon juice
- Salt and pepper to taste

Instructions:
1. Preheat oven to 400°F (200°C).
2. In a baking dish, toss asparagus with olive oil, garlic, lemon juice, salt, and pepper.
3. Roast for 15 minutes or until tender.
4. Serve warm.

Nutritional Information (per serving):
- Calories: 80
- Carbohydrates: 6g
- Protein: 3g
- Fat: 6g

Pumpkin and Ginger Soup

Prep Time: 10 minutes
Cook Time: 20 minutes
Portion Size: 2

Ingredients:
- 1 tablespoon olive oil
- 1 onion, chopped
- 2 cloves garlic, minced
- 1-inch piece ginger, minced
- 2 cups pumpkin puree
- 3 cups vegetable broth
- Salt and pepper to taste

Instructions:
1. Heat olive oil in a pot over medium heat.
2. Add onion, garlic, and ginger; sauté until softened.

3. Add pumpkin puree and vegetable broth.
4. Bring to a boil, then reduce heat and simmer for 15 minutes.
5. Blend soup until smooth using an immersion blender.
6. Season with salt and pepper to taste and serve warm.

Nutritional Information (per serving):
- Calories: 150
- Carbohydrates: 30g
- Protein: 3g
- Fat: 5g

Avocado and Spinach Green Smoothie

Prep Time: 5 minutes
Portion Size: 2

Ingredients:
- 1 avocado
- 1 cup fresh spinach
- 1 banana
- 1 cup almond milk
- 1 tablespoon honey
- 1/2 cup ice

Instructions:
1. Combine all ingredients in a blender.
2. Blend until smooth.
3. Pour into glasses and serve immediately.

Nutritional Information (per serving):
- Calories: 200
- Carbohydrates: 30g
- Protein: 3g
- Fat: 10g

Chia Seed Pudding with Coconut Milk

Prep Time: 10 minutes (plus overnight refrigeration)
Portion Size: 2

Ingredients:

- 1 cup coconut milk
- 3 tablespoons chia seeds
- 1 teaspoon vanilla extract
- 1 tablespoon honey
- 1/2 cup fresh berries (optional)

Instructions:

1. In a bowl, combine coconut milk, chia seeds, vanilla extract, and honey.
2. Stir well to combine.
3. Cover and refrigerate overnight.
4. Stir again before serving, and top with fresh berries if desired.

Nutritional Information (per serving):

- Calories: 200
- Carbohydrates: 20g
- Protein: 4g
- Fat: 12g

Apple Cider Vinegar Detox Drink

Prep Time: 5 minutes
Portion Size: 2

Ingredients:

- 2 cups water
- 2 tablespoons apple cider vinegar
- 1 tablespoon honey
- 1 teaspoon lemon juice
- 1/2 teaspoon ground cinnamon

Instructions:

1. In a pitcher, combine all ingredients.
2. Stir well to mix.
3. Pour into glasses and serve immediately.

Nutritional Information (per serving):

- Calories: 15
- Carbohydrates: 4g
- Protein: 0g
- Fat: 0g

Beet and Carrot Salad with Apple Cider Vinegar Dressing

Prep Time: 10 minutes
Portion Size: 2

Ingredients:
- 1 medium beet, grated
- 2 carrots, grated
- 2 tablespoons apple cider vinegar
- 1 tablespoon olive oil
- 1 teaspoon honey
- Salt and pepper to taste

Instructions:
1. In a bowl, combine grated beet and carrots.
2. In a small bowl, whisk together apple cider vinegar, olive oil, honey, salt, and pepper.
3. Pour dressing over beet and carrot mixture.
4. Toss to combine and serve immediately.

Nutritional Information (per serving):
- Calories: 90
- Carbohydrates: 14g
- Protein: 1g
- Fat: 4g

Quinoa Salad with Fermented Vegetables

Prep Time: 15 minutes
Cook Time: 15 minutes
Portion Size: 2

Ingredients:
- 1/2 cup quinoa, rinsed
- 1 cup water
- 1/2 cup mixed fermented vegetables (e.g., sauerkraut, kimchi)
- 1/2 cup cherry tomatoes, halved
- 1/4 cup chopped cucumber
- 2 tablespoons olive oil
- 1 tablespoon lemon juice
- Salt and pepper to taste

Instructions:

1. In a saucepan, bring quinoa and water to a boil. Reduce heat, cover, and simmer for 15 minutes or until water is absorbed.
2. In a large bowl, combine cooked quinoa, fermented vegetables, cherry tomatoes, and cucumber.
3. Drizzle with olive oil and lemon juice.
4. Toss gently to combine and season with salt and pepper to taste.

Nutritional Information (per serving):
- Calories: 200
- Carbohydrates: 28g
- Protein: 5g
- Fat: 8g

Blueberry and Almond Overnight Oats

Prep Time: 10 minutes (plus overnight refrigeration)
Portion Size: 2

Ingredients:
- 1 cup rolled oats
- 1 cup almond milk
- 1/2 cup fresh blueberries
- 2 tablespoons almond butter
- 1 tablespoon honey

Instructions:
1. In a bowl, combine rolled oats, almond milk, blueberries, almond butter, and honey.
2. Stir well to combine.
3. Cover and refrigerate overnight.
4. Stir again before serving.

Nutritional Information (per serving):
- Calories: 250
- Carbohydrates: 40g
- Protein: 7g
- Fat: 8g

Cabbage and Carrot Coleslaw with Yogurt Dressing

Prep Time: 15 minutes
Portion Size: 2

Ingredients:
- 2 cups shredded cabbage
- 1 carrot, grated
- 1/4 cup plain Greek yogurt
- 1 tablespoon apple cider vinegar
- 1 teaspoon honey
- Salt and pepper to taste

Instructions:
1. In a large bowl, combine shredded cabbage and grated carrot.
2. In a small bowl, whisk together Greek yogurt, apple cider vinegar, honey, salt, and pepper.
3. Pour dressing over cabbage and carrot mixture.
4. Toss to combine and serve immediately.

Nutritional Information (per serving):
- Calories: 70
- Carbohydrates: 12g
- Protein: 3g
- Fat: 2g

Turmeric and Ginger Infused Water

Prep Time: 5 minutes
Portion Size: 2

Ingredients:
- 4 cups water
- 1 teaspoon ground turmeric
- 1 teaspoon grated ginger
- 1 tablespoon lemon juice
- 1 teaspoon honey

Instructions:
1. In a pitcher, combine water, turmeric, ginger, lemon juice, and honey.
2. Stir well to mix.
3. Refrigerate for at least 1 hour before serving.

Nutritional Information (per serving):
- Calories: 10
- Carbohydrates: 2g
- Protein: 0g
- Fat: 0g

Garlic and Rosemary Roasted Sweet Potatoes

Prep Time: 10 minutes
Cook Time: 30 minutes
Portion Size: 2

Ingredients:
- 2 medium sweet potatoes, peeled and diced
- 1 tablespoon olive oil
- 2 cloves garlic, minced
- 1 teaspoon dried rosemary
- Salt and pepper to taste

Instructions:
1. Preheat oven to 400°F (200°C).
2. In a bowl, toss sweet potatoes with olive oil, garlic, rosemary, salt, and pepper.
3. Spread sweet potatoes on a baking sheet in a single layer.
4. Roast for 30 minutes or until tender and slightly browned.
5. Serve warm.

Nutritional Information (per serving):
- Calories: 180
- Carbohydrates: 32g
- Protein: 2g
- Fat: 6g

Probiotic Banana and Oat Muffins

Prep Time: 15 minutes
Cook Time: 20 minutes
Portion Size: 2

Ingredients:
- 1 cup rolled oats
- 1 cup plain Greek yogurt
- 2 ripe bananas, mashed
- 1/4 cup honey
- 1 teaspoon baking powder
- 1/2 teaspoon baking soda
- 1/2 teaspoon cinnamon
- 1/4 teaspoon salt

Instructions:

1. Preheat oven to 350°F (175°C).
2. In a bowl, combine all ingredients and mix well.
3. Divide batter evenly among muffin cups.
4. Bake for 20 minutes or until a toothpick inserted into the center comes out clean.
5. Let cool before serving.

Nutritional Information (per serving):
- Calories: 200
- Carbohydrates: 40g
- Protein: 6g
- Fat: 3g

Probiotics, Prebiotics, and Fermented Foods for Gut Health

Probiotics:
1. **Yogurt**: Rich in live cultures that promote a healthy gut flora.
2. **Kefir**: A fermented milk drink loaded with beneficial bacteria.
3. **Sauerkraut**: Fermented cabbage that supports gut health.
4. **Kimchi**: Spicy fermented vegetables, often including cabbage and radishes.
5. **Miso**: A fermented soybean paste used in soups and marinades.
6. **Tempeh**: A fermented soybean product rich in probiotics and protein.
7. **Natto**: Fermented soybeans, known for their high probiotic content.
8. **Kombucha**: A fermented tea drink that boosts gut health.

Prebiotics:
1. **Garlic**: Contains inulin, a type of prebiotic fiber.
2. **Onions**: High in prebiotic fibers like fructooligosaccharides (FOS).
3. **Leeks**: A great source of prebiotics, supporting beneficial bacteria growth.
4. **Asparagus**: Rich in inulin, which feeds healthy gut bacteria.
5. **Bananas**: Contains resistant starch, a type of prebiotic.
6. **Chicory Root**: Extremely high in inulin, supporting digestive health.
7. **Jerusalem Artichokes**: Also known as sunchokes, high in inulin.
8. **Barley**: Contains beta-glucan, a prebiotic fiber.

Fermented Foods:
1. **Pickles**: Naturally fermented pickles are a good source of probiotics.
2. **Fermented Carrots**: A crunchy and probiotic-rich snack.
3. **Fermented Beets**: Supports digestion and provides beneficial bacteria.
4. **Fermented Radishes**: Adds variety and probiotics to your diet.
5. **Lacto-fermented Vegetables**: Various vegetables fermented using lactobacillus bacteria.
6. **Rejuvelac**: A fermented grain drink high in enzymes and probiotics.
7. **Sourdough Bread**: Made through a fermentation process that enhances its digestibility.
8. **Apple Cider Vinegar**: Contains beneficial bacteria and acetic acid that support gut health

Conclusion

In the journey of life, our health and well-being are paramount. Through the natural remedies, wholesome recipes, and wise principles provided by Dr. Barbara O'Neill, we are given the tools to nurture and care for our bodies in profound and transformative ways. This book has been a comprehensive guide, offering natural solutions for heart health, weight management, blood sugar balance, joint health, and beyond. By embracing these remedies, we empower ourselves to achieve optimal health and vitality.

Each recipe and remedy within these pages is designed to support and enhance various aspects of your health, from strengthening your heart to balancing your hormones, boosting your brain power, and improving your gut health. These natural approaches harness the power of whole foods, herbs, and time-tested techniques to promote healing, prevent illness, and sustain a vibrant, healthy life.

As you continue on your path to wellness, remember that the choices you make every day—what you eat, how you move, and how you care for yourself—are the foundation of your health. By integrating these natural remedies into your daily routine, you take proactive steps toward a healthier, more balanced life.

Thank you for allowing this book to be a part of your journey. May the knowledge and recipes within these pages inspire you to live with vitality and joy, embracing the natural, holistic approach to health that Dr. Barbara O'Neill advocates. Here's to your health, happiness, and a future filled with well-being.